TRANS YOUR LIFE AND SAVE THE WORLD

– Through Living In Support Of The Biological Truth About The Human Condition –

Jeremy Griffith

This short book is biologist Jeremy Griffith's powerful condensation of his definitive treatise on the human condition in *FREEDOM: The End Of The Human Condition*

'THE BOOK THAT SAVES THE WORLD'
PROFESSOR HARRY PROSEN,
former president of the Canadian Psychiatric Association

FREEDOM
THE END OF THE HUMAN CONDITION

At last, the redeeming, reconciling and rehabilitating
biological explanation of the human condition that
brings about the dreamed of dawn of understanding
and ends all the suffering and conflict on Earth

JEREMY GRIFFITH
www.HumanCondition.com

www.HumanCondition.com

TRANSFORM YOUR LIFE AND SAVE THE WORLD Through Living In Support Of The Biological Truth About The Human Condition by Jeremy Griffith

First Edition, published in 2016, by WTM Publishing and Communications Pty Ltd (ACN 103 136 778) (www.wtmpublishing.com)

Cover design by James Press

All enquiries to:
WORLD TRANSFORMATION MOVEMENT® (WTM®)
GPO Box 5095, Sydney NSW 2001, Australia
Phone: + 61 2 9486 3308   Fax: + 61 2 9486 3409
Email: info@worldtransformation.com
Website: www.humancondition.com or www.worldtransformation.com

The WTM is a non-profit organisation which holds an Authority to Fundraise for Charitable Purposes in NSW, Australia.

ISBN 978-1-74129-035-6
CIP – Biology

Edited by Fiona Cullen-Ward.
Typesetting: designed by Jeremy Griffith, set by Lee Jones and Brony FitzGerald (3). Font: Times; main body text: 10.7pt on 13.37pt leading; quote: 9.21pt bold; quote source: 8.126pt; comment within a quote: 10.27pt; digits and all caps text: 1-2pts smaller than body text. For further details about the typesetting, styles and layout used in this book please view the *WTM Style Guide* at www.humancondition.com/style-guide.

Contents

Notes to the Reader

While this book can be read as a stand-alone document, its main function is to serve as the recommended introduction to Australian biologist Jeremy Griffith's definitive explanation of the human condition that is presented in his 2016 book, *FREEDOM: The End Of The Human Condition*.

As a condensation of *FREEDOM*, there are references throughout to chapters and paragraphs in *FREEDOM* where you can find more complete descriptions of the concept being discussed, including quote sources. Highlights have also been added to assist the reader.

'Part 1: The Dishonest Biology' and 'Part 2: The Truthful Biology' represent an expanded transcript of Jeremy Griffith's address at the launch of *FREEDOM* at the Royal Geographical Society (RGS) in London on 2 June 2016—a launch at which the great humanitarian and 'conscience of the world', Sir Bob Geldof, introduced Griffith. 'Part 3: The Resulting Transformation Of The Human Race' is the expanded transcript of a presentation Griffith gave on his return to Australia in August 2016.

A film of Griffith's RGS address followed by his transformation presentation can be viewed at www.HumanCondition.com/transform.

Both *FREEDOM* and *Transform Your Life And Save The World* are **freely** available online to be read, printed or shared at www.HumanCondition.com; alternatively, you can purchase copies from bookstores or Amazon. *FREEDOM* has a very informative Introduction by Professor Harry Prosen, former President of the Canadian Psychiatric Association.

The website also provides a range of helpful information, including introductory videos about the human condition, its resolution and transforming effects; a description of the World Transformation Movement and its membership; a Media Centre; and biographical information about Griffith.

www.HumanCondition.com

Introduction

This introduction consists of short extracts from the introductory presentations by Tim Macartney-Snape and Sir Bob Geldof at the launch of Jeremy Griffith's book *FREEDOM: The End Of The Human Condition* at the Royal Geographical Society (RGS) in London on 2 June 2016.

(Watch the film of the RGS launch and Griffith's subsequent transformation address at www.HumanCondition.com/transform.)

Extract from the Introduction by Tim Macartney-Snape
AM OAM (world-renowned mountaineer, biologist and twice
honoured Order of Australia recipient; Tim is also a Founding
Member, and now a Patron, of the World Transformation
Movement, the organisation established by Jeremy Griffith in
1983 for the study and amelioration of the human condition)

Welcome everyone to the launch of *FREEDOM* at the Royal
Geographical Society (RGS), especially Sir Bob Geldof, who we
are extremely honoured to have as our keynote speaker…Sir Bob
will speak first about the crisis facing the world after which Jeremy
will present *the* solution to that crisis—and I say *the* solution
because what Jeremy is going to explain is so accountable and
clarifying that it realigns our lives and fixes the world…

The most quoted author in *FREEDOM* is the great philosopher Sir Laurens van der Post and in his 1990 RGS Television
Lecture of the Year he anticipated Jeremy's book and its ultimate
exploration, which is into the human condition. So it's fitting that
we're launching here today with an honoured fellow of the RGS,
Sir Bob Geldof. Before asking Sir Bob to speak I want to play a
short extract of Sir Laurens' lecture here at the RGS in 1990: **'It's
been said that the explorers in mankind must be singularly unemployed
because there's nothing left in this world to explore. Well of course
that depends on what you mean by exploration…in the sense to which
exploration is both an exploration into the physical *and* into the spirit of
man there is a lot ahead in your keeping…there is exploration in reverse
to do. We must go sharply into reverse…we must rehabilitate ourselves.
We must get our old natural selves to join with our other conscious, wilful,
rational, scientific selves…Your job is not over.'**

(Watch & read Tim's full introduction at
www.humancondition.com/tim-launch-speech.)

Extract from the keynote address by Sir Bob Geldof
FRGS (world-renowned humanitarian and musician)

I've never felt the world more threatening, more fractious, more fissiparous, more febrile, more fucked up than it is now… If there is new thinking, or new ideas, from a scientist, from a biologist, from a man who has used his life on almost quixotic expeditions to find Tasmanian Tigers, but to prove they live or don't live, that they exist or they don't exist, to prove it, to be a scientist, then you have to pay attention to a person who thinks about what a genius like van der Post has considered which influences us, and the way we live and the way we think…

We need to think, we need new ideas, we need proselytisers, we need obsessed people, which I think Jeremy is. We need him to be questioned. We need *FREEDOM* to be argued, we need it to be read and talked about and understood. It may be right, it may be wrong. But you need someone as committed as Jeremy to trying to understand what gets us here time after time. We must be better than that…

Never before have we needed genius, magic and power more; and the word I'd add to that is thought and thinking…That is why I love the RGS, you talk to these thinking people, they may agree, they may disagree, but you know it and you'll have set some little spark alight to make them think afresh and think differently. He did it with me, I hope he does it with you right now, ladies and gentleman, Jeremy Griffith.

(Watch & read Sir Bob's full address at
www.humancondition.com/geldof-launch-speech.)

Part 1

The Dishonest Biology

Thank you Tim, my wonderful, wonderful friend on this long and very difficult journey to this day, for your kind introduction.

Most especially I want to thank Sir Bob Geldof for your inspiring and encouraging address. My gratitude to you for agreeing to help with the launch of *FREEDOM* is completely bottomless, because in my view, Sir Bob, your concern for humanity is second to none, and your preparedness to relentlessly point out the horror of our species' predicament makes you the conscience of the world. I love your prophetic honesty so much that to me it's like the coming of the summer rains that water the parched Earth! Thank you.

I would like to begin by referring to some of the amazingly honest statements Sir Bob has made about the extremely serious plight of the world.

You have just stated that you've **'never felt the world more threatening, more fractious, more fissiparous, more febrile, more fucked up than it is now.'**

And in your 1986 album with the appropriate title *Deep in the Heart of Nowhere,* you sang, **'What are we going to do because it can't go on...This is the world calling. God help us'**, and **'Searching through their sacred books for the holy grail of "why", but the total sum of knowledge knows no more than you or I.'**

At a recent conference in Montreal, you asked, **'Where is the public good in the vast inequality wrought by modern capitalism...**

which is just a euphemism for greed?', and you said, 'I've never felt greater fear and greater anger than I do today.'

In 2002 you pleaded, 'Can the void, the nothingness, that we appear to inhabit ever be filled?'

In your 2006 book *Geldof in Africa,* you wrote about 'attempting to make sense of...the scale of the horror, the utter devastation' — and referred to 'world leaders' pathetic plans' to help. And asked, 'What will ever stop the madness?'

In your brilliant W.B. Yeats documentary last month, you pleaded for Yeats' 'revolution of the mind' where 'peace, pluralism and respect' will replace hate and aggression, and you referred to 'the ludicrous notion of death or glory' practised in Ireland during the 1916 Easter Uprising.

And, finally, at the very end of *Geldof In Africa* you wrote, 'I'm sick of standing in squares and linking arms...of tear gas and pop concerts...Feed the World. Yes. For we starve for justice. We hunger for dignity. We thirst for the end of degradation and hurt. Feed the world, for we are empty of hope and too full of despair and have nothing to nourish our dreams.' Basically, what you're pleading for here is the food we really need, which is understanding of the human condition.

Yes, the *real* and ultimate frontier that we needed to explore was never outer-space but inner-space — as Sir Laurens van der Post said, 'We must go sharply into reverse...get our old natural selves to join with our other conscious, wilful, rational, scientific selves'. Again, the message is that to make any progress in our human situation we had to solve the human condition.

And as I will seek to make very clear today, unless the reconciling understanding of the human condition was found, as it now is in *FREEDOM*, the rapidly escalating pain and alienation in humans was only going to increase at such a rapid rate and to such an extent that all the suffering the human race has experienced to this point in time would pale in comparison to the amount of human suffering that was coming our way.

The fact is, what is being presented here today provides the *only* doorway out of the incredible crisis the human race is now in, so please, please listen to what I have to say. Basically I'm going to present the essence of what's in *FREEDOM*, and I might mention that it's been a 50 year journey to get to this day and this venue and this presentation in, most gratefully, the esteemed company of Sir Bob.

Sir Bob, you're a fabulous humanitarian, however, when it comes to such fundamental questions as '**the holy grail of "why"**' the human condition exists, you said to me a few days ago that '**they are questions for biologists like you Jeremy to answer**'—and on that score you remarked that '**We're not all going to turn into people who are all hugging each other Jeremy because we're all competitive by nature. The question is how do we relieve ourselves from our unchangeable primal instincts** [that are selfish and competitive]?' Essentially you were expressing the view virtually everyone has been subscribing to, which is that we humans have a competitive and aggressive animal heritage, and thus we have brutish and savage genetic instincts that we can't change. But, as I am now going to explain, that's all been a great lie, albeit a very necessary one. And after explaining why this has been a lie, I will go on to explain that our condition is a *psychological* condition, and that psychoses *can* be healed with understanding.

It is certainly true that humans have used the excuse that we are competitive, aggressive and selfish because we supposedly have instincts that compel us to compete for food, shelter, territory and a mate in order to reproduce our genes, just as other animals do—but this cannot be the *real* reason for our competitive and aggressive behaviour.

First, we humans have cooperative and loving moral instincts, the voice or expression of which within us is our conscience—as Charles Darwin recognised, '**The moral sense perpahs affords the best and highest distinction between man and the lower animals**' (paragraph 375 of *FREEDOM*). And in order to have an instinctive altruistic moral nature, our distant ancestors must have been cooperative and loving, *not* competitive and aggressive like other animals.

Second, descriptions of human behaviour, such as egocentric, arrogant, inspired, depressed, deluded, pessimistic, optimistic, artificial, hateful, mean, immoral, guilt-ridden, evil, psychotic, neurotic, alienated, etc, all recognise the involvement of our species' unique fully conscious thinking mind—that there is a *psychological* dimension to *our* behaviour. Humans have suffered *not* from the genetic-opportunism-based, non-psychological *animal* condition, but the conscious-mind-based, *PSYCHOLOGICALLY* troubled *HUMAN CONDITION*. (par. 40 of *FREEDOM*)

So saying our behaviour results from having competitive and aggressive instincts like other animals simply does not stack up, it is not true; it's just an excuse, but, again, an extremely necessary one, as I will now explain.

To start, let's imagine that it is true that our distant ancestors did live cooperatively and lovingly, the instinctive memory of which is our '**moral sense**' or conscience that Darwin recognised distinguishes us from other animals.

And following that first step, let's imagine that Plato was right when, long ago, in 360 BC he wrote of '**our state of innocence, before we had any experience of evils to come, when we were…simple and calm and happy…pure ourselves and not yet enshrined in that living tomb which we carry about, now that we are imprisoned**', a time when we lived a '**blessed and spontaneous life…**[where] **neither was there any violence, or devouring of one another** [no sex as humans practice it now]**, or war or quarrel among them…And they dwelt naked, and mostly in the open air…and they had no beds, but lay on soft couches of grass**' (pars 158 & 170 of *FREEDOM*).

Similarly, let's imagine that Plato's compatriot Hesiod was right when, in his poem *Theogony*, he wrote, '**When gods alike and mortals rose to birth / A golden race the immortals formed on earth... Like gods they lived, with calm untroubled mind / Free from the toils and anguish of our kind...Strangers to ill, their lives in feasts flowed by...They with abundant goods 'midst quiet lands / All willing shared the gathering of their hands'** (par. 180).

And let's imagine that our mythologies were also right; that, as the author Richard Heinberg summarised, '**Every religion begins with the recognition that human consciousness has been separated from the divine Source, that a former sense of oneness...has been lost... everywhere in religion and myth there is an acknowledgment that we have departed from an original...innocence and can return to it only through the resolution of some profound inner discord...the cause of the Fall is described variously as disobedience, as the eating of a forbidden fruit** [from the tree of knowledge], **and as spiritual amnesia** [forgetting, blocking out, alienation]' (par. 181).

And let's then imagine that the philosopher Jean-Jacques Rousseau was right when he said about even some humans living today that '**nothing is more gentle than man in his primitive state**' (par. 181)—and that the anthropologist C. Owen Lovejoy was correct in his summary of the evidence that anthropology has now revealed, which is that '**our species-defining cooperative mutualism can now be seen to extend well beyond the deepest Pliocene** [well beyond 5.3 million years ago]' (par. 183).

And, from there, let's imagine that it is *also* true that primatological studies of bonobos (the variety of chimpanzee living south of the Congo river) reveal a great deal about how loving and cooperative our distant primate ancestors likely were.

To evidence this last point, consider these absolutely amazing observations of bonobos. Firstly, from filmmakers who were producing a documentary about them: '**they're surely the most fascinating animals on the planet. They're the closest animals to man.**

They're the only animals capable of creating the same "gaze" as a human.'
And 'Once I got hit on the head with a branch that had a bonobo on it. I
sat down and the bonobo noticed I was in a difficult situation and came
and took me by the hand and moved my hair back, like they do. So they
live on compassion, and that's really interesting to experience' (par. 418).

The bonobo Kanzi's human-like gaze (from paragraph 418)

Yes, as bonobo keeper Barbara Bell said, 'Adult bonobos
demonstrate tremendous compassion for each other...For example, Kitty,
the eldest female, is completely blind and hard of hearing. Sometimes she
gets lost and confused. They'll just pick her up and take her to where she
needs to go.' And that 'They're extremely intelligent...They understand
a couple of hundred words...It's like being with 9 two and a half year
olds all day' and 'They also love to tease me a lot...Like during training,
if I were to ask for their left foot, they'll give me their right, and laugh
and laugh and laugh' (par. 451).

Researchers have also reported that **'bonobos historically have existed in a stable environment rich in sources of food…and unlike chimpanzees have developed a more cohesive social structure'**. For example, **'up to 100 bonobos at a time from several groups spend their night together. That would not be possible with chimpanzees because there would be brutal fighting between rival groups.'** (par. 415)

Bonobo group (from paragraph 415)

Note that what's being said about bonobos is amazingly similar—well, it's almost identical—to how Plato described our distant ancestors long before bonobos were even known of, which was that they **'dwelt naked, and mostly in the open air…and they had no beds, but lay on soft couches of grass'**. And, in Hesiod's words, **'They with abundant goods 'midst quiet lands / All willing shared the gathering of their hands.'**

Bonobos' unlimited capacity for love is also apparent in this truly amazing first-hand account from researcher Vanessa Woods: **'Bonobo love is like a laser beam. They stop. They stare at you as though they have been waiting their whole lives for you to walk into their jungle. And then they love you with such helpless abandon that you love them back. You have to love them back'** (par. 451).

And significantly, unlike other primate societies, bonobo society is matriarchal and focused on the nurturing of their infants, as this quote evidences: **'Bonobo life is centered around the offspring. Unlike what happens among chimpanzees, all members of the bonobo social group help with infant care and share food with infants. If you are a bonobo infant, you can do no wrong...Bonobo females and their infants form the core of the group'** (par. 416).

Frans Lanting/Mint Images/Getty Images; San Diego Zoo (bottom right)

Bonobos nurturing their infants (paragraph 417)

All these descriptions (and you can read many more in chapter 5 of *FREEDOM*) provide powerful insights into how amazingly cooperative and loving bonobos, our closest living relative, are. In fact, the following picture shows just how similar our species are, comparing as it does the skeleton of our early australopithecine ancestor (who lived between 3.9 and 3 million years ago) with the skeleton of a bonobo (chs 8:2 and 8:3).

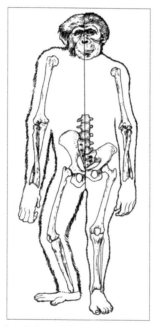

Left side: Bonobo skeleton. Right side: Early australopithecine.
(Drawing by Adrienne L. Zihlman from *New Scientist*, 1984)

On the back of those primatological insights, let's now imagine that the American philosopher John Fiske was right when, in 1874 (only a few years after Darwin published his great book *Origins*), he proposed that it was through the nurturing of our infants that our distant primate ancestors acquired co-operative loving moral instincts (nurturing like that the bonobos practise). As I explain in chapter 5 of *FREEDOM*, the reason nurturing can create a cooperative moral sense is that while the trait for nurturing is a selfish trait, as genetic traits have to be to survive, it has the *appearance* of being selfless. While the mother is selfishly reproducing her genes by fostering her infant, from the infant's perspective, it is being treated unconditionally selflessly—the mother is giving her offspring food, warmth, shelter, support and protection for *apparently* nothing in return. So it follows that if the infant can remain in infancy for an extended period and be treated with a lot of seemingly altruistic love, they will be indoctrinated with that selfless love and grow up to behave accordingly. Of course, being semi-upright from living in trees, and thus having their arms free to hold a dependent infant, primates are especially facilitated to develop this nurtured, loving, cooperative nature. So yes, let's imagine that it was entirely appropriate that Fiske's nurturing theory for the origin of our moral nature was described at the time as being **'far more important'** than **'Darwin's principle of natural selection'** and **'one of the most beautiful contributions ever made to the Evolution of Man'** (par. 488).

With all this in mind, let's now imagine that it was the emergence of our conscious mind (which, as explained in chapter 7 of *FREEDOM*, an environment of unconditional selflessness or love liberated the development of) that caused our species to depart from that original all-loving, all-sensitive, completely happy and cooperative idyllic state—that it was, as Heinberg wrote, **'human consciousness'** that **'separated'** us **'from an original... innocence'**. And let us imagine therefore that Shakespeare was right when he acknowledged how seemingly destructive and stupid we marvellously conscious humans are, writing **'What a piece of work is a man! how noble in reason! how infinite in faculty!... in action how like an angel! in apprehension how like a god! the beauty of the world! the paragon of animals! And yet, to me, what is this quintessence of dust? [Brutal and barbaric] man delights not me.'**

And that the philosopher Blaise Pascal was similarly precise in his damning observation, **'What a chimera then is man! What a novelty, what a monster, what a chaos, what a contradiction, what a prodigy! Judge of all things, imbecile worm of the earth, repository of truth, a sewer of uncertainty and error, the glory and the scum of the universe!'** (par. 49)

Shakespeare's and Pascal's denunciations of our conscious brain certainly leave no doubt as to their views on the difficulty of understanding the contradictions faced by the conscious mind, so let's next imagine that science writer Roger Lewin was also right when, in commenting about how depressing and confronting the subject of consciousness is, he said that to **'illuminate the phenomena of consciousness'** is **'a tough challenge...perhaps the toughest of all'**. And that philosopher René Descartes' reaction when he too tried to **'contemplate consciousness'**—that **'So serious are the doubts into which I have been thrown...that I can neither put them out of my mind nor see any way of resolving them. It feels as if I have fallen unexpectedly into a deep whirlpool** [of utter depression]**'**—was completely understandable given what a destructive force consciousness has appeared to be. (par. 624)

Yes, when taking into account the cooperative and loving nature of our species' instinctive past, how ashamed of our conscious mind's seeming destruction of that state must we be!

Now, if *all* these imaginings are right (and I suggest that by the end of this presentation humans' conscious mind will be so well defended, yes *defended*, by the true explanation for our corrupted state that you will be able to admit that these imaginings are, in fact, all true), namely that our distant ape ancestor lived completely cooperatively and lovingly and then we became fully conscious and for some reason, unbeknownst to ourselves, we began to behave competitively, aggressively and selfishly, then how absolutely terrible a predicament must that have been!

Instead of being loving, cooperative and gentle, humans had developed a propensity for shocking deeds of violence, depravity, indifference and cruelty. For some reason we had turned into seemingly evil monsters—*sufferers of the human condition*. And having become preoccupied seeking power, fame, fortune and glory to salve the shame and insecurity caused by that condition meant there was little or no time left to properly nurture our children, which meant *they* in turn became victims of our corrupted condition, and so on, through generation after generation—the guilt of which explains why John Fiske's explanation of the nurturing origins of our moral nature has been completely ignored and forgotten, despite it being an insight that was described at the time as being **'far more important than Darwin's'**. It has been said that parents would rather admit to being an axe murderer than a bad mother or father, and it's true: the importance of nurturing in the development of humanity and in our own lives is a truth we have not been prepared to admit. (ch. 6:4)

We can now see that it makes complete sense that the guilt associated with trying to think about and confront the truth of the horror of our corrupted condition has been so great that we could only cope with it by not coping with it, by adopting all manner of dishonest denial—basically by blocking out all memory of our species' cooperative and loving past and any thoughts about how corrupted we as a species had become since we gained a conscious, thinking mind. We didn't want to admit our species once lived cooperatively and lovingly, and we didn't want to face the issue of the human condition. Plato's famous description of **'our human condition'** in which he depicted humans as having to **'take refuge'** **'a long way underground'** in a dark **'cave'** because they couldn't face the **'painful'** **'light'** of the truth of **'the imperfections of human life'**, is, in fact, a deadly accurate portrayal of this inability to cope with our species' corrupted reality—of how deeply resigned we humans have become to living in denial of the issue of the human condition. (pars 81-84)

Computer graphic by Jeremy Griffith, Marcus Rowell and Genevieve Salter © 2009 Fedmex Pty Ltd

Author's drawing of humans finally being able to escape Plato's metaphorical dark cave of denial into the healing sunshine of understanding of the human condition

To further explain this all-significant process of **Resignation**: between the ages of around 12 to 15 adolescents try to confront the human condition; they question and try to think truthfully about the seeming complete wrongness of human life, how it is all completely contradictory to what our original instinctive moral self or soul expects. But by the time they reach 14 or 15 their thoughts become so depressing that they learn that they have no choice but to resign themselves to never trying to confront the issue of the human condition again. In this respect, the individual goes on the same journey as the conscious-mind-based human race, the principle of biology that 'ontogeny recapitulates phylogeny'. Humanity started adopting denial to cope, and it was a strategy that each generation of humans *also* had to learn to adopt on a personal level.

It is pertinent here to include my 'valleys drawing' that depicts humanity's (and the individual's) journey from ignorance to enlightenment (this drawing appears a number of times in *FREEDOM*). At the start of the frame you can see humanity, or in the individual's case, the young adolescent, leaving the valley of our/their innocent childhood and having to resign to an absolutely horrible life of living with the upset state of the human condition, which demands you deny/forget that idyllic soulful past and the issue it raises of the human condition. This stage of Resignation, which historically has been too unbearable to acknowledge once that transition has occurred, is fully explained in chapter 2:2 of *FREEDOM* where you'll find some excruciatingly honest accounts from adolescents negotiating this horror.

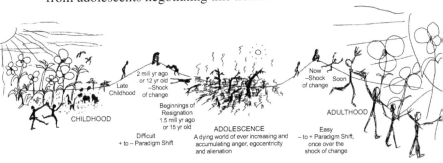

Humanity's Journey from Ignorance to Enlightenment

I might mention here that occasionally some adolescents refuse to properly resign and take up the strategy of denial of the issue of the human condition. I call these semi-unresigned individuals 'ships at sea' (par. 1209) because they refuse to pull into port, as it were, and escape the storms that occur when they try to grapple with the human condition during their early adolescence. I'm certain that Sir Bob's extraordinary capacity for honesty about the extent and horror of the human condition in the world is because he is, to a degree, a 'ship at sea'. Reading his 1986 book *Is That It?* is like reading one long 'Resignation poem', a practise many adolescents engage in during Resignation and one I refer to in chapter 2:2 of *FREEDOM*. Sir Bob repeatedly writes about **'a civil war inside my head which accelerates until I think my brain will explode'**, which is a classic description of a 'still-wrestling-with-the-human-condition'-unresigned mind. As you have said Sir Bob, **'I will always rail against those things I abhor. I will always try and avoid the cant and hypocrisy I loathe so much. I will continue being an "awkward bugger".'** Yes, truthful, unresigned people make the world of the resigned feel incredibly awkward, but thank God for such truthsayers. The only reason I can explain the human condition, as I'm about to do, is because I have an unresigned mind that still confronts and thinks truthfully about the human condition—which is why I so love Sir Bob's honesty about the state of the world.

Of course, for resigned minds that are committed to avoiding the issue of the human condition the resulting human-condition-confronting truthfulness of *FREEDOM* makes it a difficult book to read. In fact, as I describe in chapter 1:4 of *FREEDOM*, resigned minds are *so* trained to avoid thinking about the human condition that when they start reading my work they suffer from a 'deaf effect' where their minds initially find it very difficult taking in or 'hearing' what's being said, to the extent that they even end up thinking their incomprehension must be due to poor expression, or that the treatise is too dense to understand or even completely meaningless! Conversely, unresigned minds don't have to patiently allow their minds the time to overcome the deaf effect when reading my work; the blocks are not in place, and so they can 'hear' its truthful explanation of the human condition with such ease it is instantly life-saving, as this typical response from a 'ship at sea' that we received this week shows: **'Have watched the video presentations** [on your website] **and have read to part-way through chapter two** [of *FREEDOM*] **when I read the line; "To be confused and frightened to the point of being sickened by human behaviour, indeed to be suicidally depressed by it, is the effect the human condition has if you haven't resigned yourself to living a**

relieving but utterly dishonest and superficial life in denial of it." This has shaken me to the core, as it describes my tormented life of 67 years, and it tells me why it has been so, which nothing or no-one has been able to. Having discovered this immense truth, I hope I can find in the rest of this book a way to cope with the avalanche of emotions it has unleashed. At the moment I feel overwhelmed and vulnerable, which I presume is not unknown to yourselves when this realisation occurs in people like me, and would dearly love to receive some guidance on this. I honestly believe that if anything can save the world, this book has the potential to; if the world can cope with it. Thank you. Bill [real name withheld]'.
We will counsel 'Bill' to be patient as a lifetime of confusion unravels itself, and we will invite him to join the 'Former Ships at Sea Club' in which he can share his experiences. We will also explain to him that chapter 9 of *FREEDOM* (a summary of which appears as Part 3 of this book) describes how the world *can* cope with all the confronting truth about the human condition.

So, the situation was that until such time that we had the understanding (the enlightenment) we required—namely the fully accountable, truthful explanation of human behaviour that will shortly be presented in Part 2—virtually all humans had no choice but to sensibly resign themselves to not trying to confront the suicidally depressing issue of the human condition. However, in the face of such pain and condemnation, denial alone wasn't enough. What we desperately needed to come up with was an *excuse* for our corrupted condition that would relieve ourselves of the excoriating and suicidally depressing truth that our conscious mind had seemingly wrecked paradise.

And so it was at this point that we zeroed in on the afore-mentioned relieving excuse for our behaviour, namely that other animals are always fighting and competing and tearing each others throats out, and so '*That's* why we're competitive and aggressive—*we* too have savage animal instincts that compel us to compete for food, shelter, territory and a mate'. Despite knowing full well that we have a cooperative and loving moral instinctive nature that can only have been derived from a time when we lived cooperatively and lovingly, and therefore that it's patently untrue that we have a savage competitive and aggressive ancestry, our fear of suicidal depression was so great that we gratefully and determinedly embraced the lie that our distant ancestors were brutish and aggressive. Further, we whole-heartedly embraced the idea that the task *then* of our uniquely conscious mind was to try to *control* these supposed brutal, savage instincts within us.

It was an <u>absolutely brilliant excuse</u> because instead of our instincts being all-loving and thus unbearably condemning of our present non-loving state, they were made out to be vicious and brutal; *and*, instead of our conscious mind being the cause of our corruption, it was made out to be the blameless mediating 'hero' that had to manage those supposed vicious instincts within us! Rather than our instincts being good and our conscious mind bad, our instincts were made out to be bad and our intellect to be good, which, while fabulously relieving, was in-effect a complete reverse of the truth lie! (par. 153)

This excuse that we have competitive, brutish and savage animal instincts gained further momentum when it was given a biological base through the misrepresentation of Charles Darwin's idea of 'natural selection' as being a 'survival of the fittest' process. Natural selection is the process whereby some members of a population reproduce more than others in a given environment, and in the first edition of *The Origin of Species* Darwin rightfully left it undecided as to whether those individuals that reproduced more could be viewed as winners, as being 'fitter'. In later editions, however, Darwin's associates, Herbert Spencer and Alfred Wallace, persuaded him to substitute the term 'natural selection' with the term 'survival of the fittest', which is the defence most people today use to justify their competitive, selfish and aggressive behaviour: 'I'm just making sure I'm a winner!' (ch. 2:9). (The reason it was 'right' that Darwin originally left it undecided as to whether selfishness is meaningful or not is explained in chapter 4 (in particular par. 358) of *FREEDOM*. Basically, nature, like everything else in the universe, is concerned with developing the holistic, teleological, negative-entropy-driven order of matter, which is a cooperative, integrative process, not a selfish, divisive one.)

The problem that emerged with this corruption of Darwin's idea (with what became known as Social Darwinism) was that it couldn't explain instances of selfless cooperation in nature, such as that which apparently occurs in ant colonies, or when humans behave in accordance with our moral conscience and help others in need. So to circumvent this problem, human-condition-avoiding, so-called 'mechanistic' biologists created the theory of Evolutionary Psychology/Sociobiology that claimed the apparent instances of selflessness in nature, such as that seen in ant colonies, was really a case of those animals indirectly selfishly fostering their own genes by fostering the survival of their relatives' genes. In the case of ants, this explanation holds true, but when used to explain *our* moral nature, it too was found wanting; as the journalist Bryan Appleyard pointed out, biologists **'still have a gaping hole in an attempt to explain altruism. If, for example, I help a blind man cross the street, it is plainly unlikely that I am being prompted to do this because he is a close relation and bears my genes.'** (ch. 2:10)

To contrive a solution to *this* problem, the Multilevel Selection Theory was created. Its proponents (the most prominent of whom is Harvard biologist E.O. Wilson) claimed that humans have some selfish instincts selected at the individual level and some selfless instincts selected at the group level. Although this idea of 'Group Selection' giving rise to selfless instincts has been fully discredited on the grounds that any selflessness that develops will always be exploited by cheaters ('By all means, help me reproduce *my* genes but I'm not about to help you reproduce yours'), by accepting the Multilevel Selection Theory human-condition-avoiding, mechanistic, reductionist biologists have nevertheless been adopting it. (ch. 2:11)

Overall, while the Multilevel Selection Theory added

unconditionally selfless instincts to selfish instincts in the mix of what allegedly forms our species' instinctive make-up (thus countering Evolutionary Psychology's offensive denial of the fact that we have altruistic moral instincts), in essence, it still amounted to a continuation of the same old reverse-of-the-truth, 'escape-rather-than-confront-the-human-condition' agenda—that humans have villainous selfish instincts and a blameless conscious mind that has to 'step-in' to control them. (par. 210)

One big problem with this 'we-are-selfish-because-we-have-selfish-instincts' excuse—which **Sir Bob** referred to—is that it implies humans are intrinsically imperfectable, that we can't change our selfish and competitive primal nature because it's in our genes and instinctive. But as is about to be explained, our instincts *are* to be cooperative and loving and it *was* a conscious mind-based psychosis that caused us to behave competitively, aggressively and selfishly—and since psychoses *can* be healed with understanding, we *can* be cured of our competitive, aggressive and selfish behaviour. Humans are *not* intrinsically imperfectable. (par. 218)

Tragically, however, the whole of biological science is now based on this fundamental lie that 'we have competitive and aggressive instincts'. Libraries are full of this dishonest nonsense. Afraid of being holistic and teleological and confronting the human condition, science has instead been evasively mechanistic and reductionist. (ch. 2:4) As **Charles Birch**, my Templeton Prize-winning Professor of Zoology at Sydney University when I was a student there, has said, **'science can't deal with subjectivity** [the human condition]**...what we were all taught in universities is pretty much a dead end'**, and **'Biology has not made any real advance since Darwin'** (par. 225). The polymath **Arthur Koestler** similarly recognised that **'the citadel they** [mechanistic scientists] **are defending lies in ruins'** (par. 223).

The overall point is that when **Sir Bob** subscribed to the 'selfish-instincts-in-humans' account he was only doing what virtually every other human has been doing. Unfortunately, however,

while it's *fully* understandable that we humans *had to* invent—and
then resign ourselves to believing in—an excuse to cope with our
diabolically corrupted condition, it means biology, and the human
race that depends on it for clarifying understanding of our behaviour, is now festering in a cesspit of human-condition-avoiding,
dishonest biological crap; it is living **'deep underground'** in Plato's
metaphorical **'cave'** of darkness.

In fact, if the redeeming and psychologically rehabilitating,
teleological, non-mechanistic true explanation of the human condition that I'm about to present hadn't arrived to liberate humans from
Plato's **'cave'** of terrible darkness from practising so much dishonest
denial, the alienation or separation from our true self that results
from that denial would only have continued to increase, so that
very soon the human race would have succumbed to an absolutely
horrific psychotic state of terminal alienation. Indeed, the recent
flood of movies and documentaries based on zombie, apocalyptic,
escape-to-another-planet, 'we are being attacked by aliens [by our
own alienation]', doomsday-preparation, 'we need a super hero to
save the world' and other judgment-day-and-anxious-Bible-related-
epic themes—*and* the actual breakout of such extreme psychosis
that some people are led to fly hijacked planes into buildings, and
drive trucks into crowds of people, and fire bullets into anyone
they can corner, *and* such overall psychological fear in young
people about the incredibly corrupted state of the world that there
is now an epidemic of children suffering from the highly anxious
and agitated Attention Deficit Hyperactivity Disorder (ADHD),
and the even more extreme, can't-cope, completely detached state
of autism—*all* reflect the fact that the end play state of terminal
alienation for the human race *is* upon us. (ch. 8:16G)

I should reinforce to the reader here that living in Plato's
cave of darkness from having to resign to a life lived in denial of
the suicidally depressing issue of the human condition, and the
truth that our species' once lived in a cooperative and loving state,
has rendered resigned adults unable to recognise how corrupted

from that original all-loving state our world has become. They find it difficult to see how psychotic/soul-separated/alienated/sick/ unwell the human race now is, and in their blindness carry on as if everyone is reasonably happy and healthy, and that life for humans today is more or less as it should be. Children, however, being unresigned, can still see and somehow have to cope with the horror of the real extent of that corruption. While it is true that the 'cave-dwelling' resigned world of denial is being forced to recognise that there are serious crises occurring from unsustainable levels and amounts of environmental devastation, resource depletion, over-population, wars, terrorism, refugees, polarised politics, corruption, failed economies, family breakdown and even mental illness, it has not been able to recognise that the *real, underlying, source* threat humankind faces is from the extremely rapidly increasing levels of alienation from our species' original innocent, all-loving state. The resigned world can't see that our species is about to die from terminal psychosis, but the fact is, the psychosis-relieving understanding of the human condition that I'm about to present has arrived in what are the absolute 'last seconds' of the metaphorical 'eleventh hour' for the human race! (par. 237)

In a further example of the blindness of the resigned mind, the failure to acknowledge that our species did once live in a cooperative and loving state means there has been a subsequent failure to see how the recent enormous advances in communication technology have destroyed the trust we derive from our species' original instinctive inclination to believe in the goodness of others; to believe that others will behave lovingly and co-operatively. There has been a failure to see that over-exposure to upset behaviour has eliminated naivety from Earth—that too much exposure to corruption has meant our soul has lost its power to cause people to behave cooperatively and lovingly. There has been an inability to recognise that cynicism about the goodness of humans is now so rampant that greed resulting from the view that 'If you're not putting yourself first you're a sucker' is now endemic—which is leading to the complete breakdown of society and to levels of soul-death/alienation escalating by the minute everywhere! The situation is so dire that if we were to measure and graph the rate of increase in alienation since the 1950s, that graph would now be climbing vertically! (pars 1026-1027)

Yes, the truth is that if the relieving understanding of the human condition that is about to be presented here hadn't been found then unimaginable and unthinkable terminal levels of writhing and deadening alienation/psychosis and the extinction of humankind were just around the corner!

Given how evasive, truth-avoiding, 'everything-is-fine-we-are-doing-well', *massively* deluded, the resigned mind is, it is really to an unresigned adolescent mind that we should go for a truly accurate description of the seriousness of our species' plight to conclude Part 1—and that's exactly what we have in these clearly unresigned, denial-free, honest lyrics from the 2010 *Grievances* album of the young American heavy metal band With Life In Mind: **'It scares me to death to think of what I have become... I feel so lost in this world'**, **'Our innocence is lost'**, **'I scream to the sky but my words get lost along the way. I can't express all the hate that's led me here and all the filth that swallows us whole. I don't want to be part of all this insanity. Famine and death. Pestilence and war.** [Famine, death, pestilence and war are traditional interpretations of the 'Four Horsemen of the Apocalypse' described in Revelation 6 in the Bible. Christ referred to similar **'Signs of the End of the Age'** (Matt. 24:6-8 and Luke 21:10-11).] **A world shrouded in darkness...Fear is driven into our minds everywhere we look'**, **'Trying so hard for a life with such little purpose...Lost in oblivion'**, **'Everything you've been told has been a lie...We've all been asleep since the beginning of time. Why are we so scared to use our minds?'**, **'Keep pretending; soon enough things will crumble to the ground...If they could only see the truth they would coil in disgust'**, **'How do we save ourselves from this misery...So desperate for the answers...We're straining on the last bit of hope we have left. No one hears our cries. And no one sees us screaming'**, **'This is the end.'** (par. 229)

Saying, '**I scream to the sky but my words get lost along the way. I can't express all the hate that's led me here and all the filth that swallows us whole. I don't want to be part of all this insanity. Famine and death. Pestilence and war**', all of which is happening in our *supposedly* normal and reasonably healthy and happy world, is exactly what Edvard Munch's famous painting *The Scream* captures. And saying, '**Everything you've been told has been a lie**' reiterates the extent of the dishonest denial in the world, especially in science, today. And saying, '**So desperate for the answers**' confirms how incredibly important are the '**answers**' about our human condition that are now going to be presented.

Edvard Munch's *The Scream* 1895

After sending this book to the printers, an article from *The Wall Street Journal* by Ed Ballard (23 Aug. 2016) was re-printed in *The Australian* with the title 'Yeats's words suit our grim times' (see <www.wtmsources.com/197>). Due to its confirmation of all that has just been said, we delayed the printing to allow this reference to be added. Employing the same words as I have, the article is about the now widespread **'feeling that something appalling is around the corner'** for humanity. As evidence of that fear, Ballard cites the recent frequency of references in the media to the chilling words I've underlined from WB Yeats' prophetic 1919 poem *The Second Coming*: **'Turning and turning in the widening gyre / The falcon cannot hear the falconer; / Things fall apart; the centre cannot hold; / Mere anarchy is loosed upon the world, / The blood-dimmed tide is loosed, and everywhere / The ceremony of innocence is drowned; / The best lack all conviction, while the worst / Are full of passionate intensity. // Surely some revelation is at hand; / Surely the Second Coming is at hand. / The Second Coming! Hardly are those words out / When a vast image out of Spiritus Mundi / Troubles my sight: somewhere in sands of the desert; / A shape with lion body and the head of a man, / A gaze blank and pitiless as the sun, / Is moving its slow thighs, while all about it / Reel shadows of the indignant desert birds. / The darkness drops again; but now I know / That twenty centuries of stony sleep / Were vexed to nightmare by a rocking cradle, / And what rough beast, its hour come round at last, / Slouches towards Bethlehem to be born?'** To mention just one of Ballard's statistics, in the first 7 months of this year the line **'The centre cannot hold'** has been used 249 times in newspapers, news sites and blogs, a figure already level with last year's record total. Note how **'twenty centuries of stony sleep'** echoes With Life In Mind's **'We've all been asleep since the beginning of time'**. The latter part of Yeats' poem is about the subconscious **'nightmare'** fear all humans have of the arrival of the **'revelation'**, judgment-day, all-exposing **'pitiless as the sun'**, **'rough [hurtful] beast'** of the truth about their massively corrupted condition, an unjustified fear that is dealt with in Part 3. The true, contexted meaning of **'the Second Coming'** is given in par. 1278 of *FREEDOM*.

Part 2

The Truthful Biology

What I will now present is the fully accountable, *truthful* biological explanation of human behaviour, which is outlined in chapter 1, and presented in detail in chapter 3, of *FREEDOM*. This is the critically important part of my presentation today because it presents the understanding that saves our species from imminent 'appalling' terminal psychosis and extinction.

Clearly what's been needed to haul the human race out of this cesspit of deceit and the psychological trauma it's creating—this deathly dark cave of denial the human race (including its vehicle for enquiry, science) has been living in and the extreme and terrible psychosis that such disconnection from our true selves is causing—is a completely fresh approach to biological thinking that truthfully recognises the association between the emergence of consciousness in humans and the corruption of our original cooperative, all-loving instinctive state. And, in fact, when that honest approach is taken the fully accountable, true biological explanation of our competitive, selfish and aggressive behaviour becomes reasonably obvious—*because it makes sense that when our species became fully conscious a battle must have broken out between it and our already established instinctive self and that this internal conflict caused us to become angry, egocentric and alienated; to become psychologically upset sufferers of the human condition.*

Even if we put aside the fact that we have cooperative loving instincts and just look at the fundamental situation an animal species would face if it developed a fully conscious mind in the presence of already established instinctive orientations, we can see how a psychologically upsetting conflict between those already established instinctive orientations, and its newly emerged, self-adjusting conscious mind would have to occur.

To help visualise this development, consider the situation of a migrating bird that has just acquired a fully conscious mind.

Many bird species are perfectly orientated to instinctive migratory flight paths. Each winter, without ever 'learning' where to go and without knowing why, they quit their established breeding grounds and migrate to warmer feeding grounds. They then return each summer and so the cycle continues. Over the course of thousands of generations and migratory movements, only those birds that happened to have a genetic make-up that inclined them to follow the right route survived. Thus, through natural selection, they acquired their instinctive orientation.

So imagine a flock of migrating storks returning to their summer breeding nests on the rooftops of Europe from their winter feeding grounds in the swamps of southern Africa. Suppose in the instinct-controlled brain of one of them we place a fully conscious mind (we will call the stork Adam because we will soon see that, up to a point, this analogy parallels the old, pre-scientific Biblical account of Adam and Eve taking the **'fruit' 'from the tree of...knowledge'**; that is, becoming conscious). As Adam Stork flies north he spots an island off to the left with a tree laden with apples. Using his newly acquired conscious mind, Adam thinks, 'I should fly down and eat some apples.' It seems a reasonable thought but he can't know if it is a good decision or not until he acts on it. For Adam's new thinking mind to make sense of the world he has to learn by trial and error and so he decides to carry out his first grand experiment in self-management by flying down to the island and sampling the apples.

The Story of Adam Stork

INTELLECT	INSTINCT
evil	good
guilty	innocent
corrupted	pure
immoral	moral
selfish	selfless
aggressive	loving
competitive	cooperative
divisive	integrative
'unGodly'	'Godly'
egotistical	altruistic
realism	idealism
conscious	conscience
reason	emotion
mind	soul
'I think'	'I feel'
head	heart
wisdom	ignorance
science	religion
mechanism	holism
right-wing	left-wing
capitalism	socialism
materialism	spiritualism
old	young
men	women
whites	blacks
artificial	natural
alienated	sound
dishonest	honest
insecure	secure
troubled	happy
city	country
sexual	non-sexual
yang	yin
dark	light

GOODIE

BADDIE

Drawing by Jeremy Griffith © 2016 Fedmex Pty Ltd

But it's not that simple. As soon as Adam's conscious thinking self deviates from his established migratory path, his innocent instinctive self (innocent in the sense of being unaware or ignorant of the need to search for knowledge) tries to pull him back on course. In following the flight path past the island, Adam's instinct-obedient self is, in effect, criticising his conscious mind's decision to veer off that course; it is condemning his search for understanding. All of a sudden Adam is in a dilemma: if he adheres to the route his instinctive self is following and flies back on course, his instincts will be happy but he'll never *learn* if his deviation was the right decision or not. All the messages he's receiving from within inform him that obeying his instincts is good, is right, but there's also a new inclination to disobey, a defiance of instinct. Diverting from his course will result in apples and understanding, yet he already sees that doing so will make him feel bad.

Uncomfortable with the criticism his newly conscious mind or intellect is receiving from his instinctive self, Adam's first response is to ignore the temptation the apples present and fly back on course. As he does, however, Adam realises he can't deny his now questioning intellect—sooner or later he must find the courage to master his conscious mind by carrying out experiments in understanding. So, continuing to think, he then asks himself, 'Why not fly down to an island and rest?' Again, not knowing any reason why he shouldn't, Adam proceeds with his experiment. And *again*, his decision is met with the same criticism from his instinctive self—but *this time* Adam *defies* the criticism and perseveres with his experimentation in self-management. His decision, however, means he must now live with the criticism and immediately he is condemned to a state of psychological 'upset'. A battle has broken out between his instinctive self, which is perfectly orientated to the flight path, and his emerging conscious mind, which needs to understand *why* that flight path is the correct course to follow. His instinctive self is perfectly orientated, but Adam doesn't *understand* that orientation, and until he does he is condemned to a divisive existence.

In short, when the fully conscious mind emerged it wasn't enough for it to be orientated by instincts, it *had to* find understanding to operate effectively and fulfil its great potential to manage life. But, tragically, the instinctive self didn't 'appreciate' that need and 'tried to stop' the mind's necessary search for knowledge, as represented by the latter's experiments in self-management—hence the ensuing battle between instinct and intellect. To refute the criticism from his instinctive self, Adam needed the discoveries that science has only recently given us of the difference in the way genes and nerves process information; in particular, he needed to be able to explain that the gene-based learning system can orientate species to situations but is incapable of insight into the nature of change.

Genetic selection of one reproducing individual over another reproducing individual (the selection, in effect, of one idea over another idea, or one piece of information over another piece of information) gives species adaptations or orientations—instinctive programming—for managing life, but those genetic orientations, those instincts, are not understandings. The nerve-based learning system on the other hand, can, if sufficiently developed, *understand* change. Nerves were originally developed for the coordination of movement in animals, but, once developed, their ability to store impressions—what we refer to as 'memory'—gave rise to the potential to develop understanding of cause and effect. If you can remember past events, you can compare them with current events and identify regularly occurring experiences. This knowledge of, or insight into, what has commonly occurred in the past enables you to predict what is likely to happen in the future and to adjust your behaviour accordingly. Once insights into the nature of change are put into effect, the self-modified behaviour starts to provide feedback, refining the insights further. Predictions are compared with outcomes and so on. Much developed, nerves can sufficiently *associate* information to *reason* how experiences are related, learn to *understand* and become *conscious* of, or aware of, or *intelligent* about, the relationship between events that occur through time. Thus consciousness means being sufficiently aware of how experiences are related to attempt to manage change from a basis of understanding. (Again, the Adam Stork story and related aspects are fully described in chapters 1 and 3 of *FREEDOM*.)

What this means is that when the nerve-based learning system became sufficiently developed for consciousness to emerge and with it the ability to understand the world, it wasn't enough to be instinctively *orientated* to the world, conscious *understanding* of the world had to be found. The problem, of course, was that Adam had only just taken his first, tentative steps in the search for knowledge, and so had no ability to explain anything. It was a catch-22 situation for the fledgling thinker, because in order to explain himself he needed the very knowledge he was setting out to accumulate. He had to search for understanding, ultimately self-understanding, understanding of why he had to 'fly off course', without the ability to first explain why he needed to 'fly off course'. And without that defence, he had to live with the criticism from his instinctive self and was *INSECURE* in its presence.

It was an untenable position to maintain and so to resist the tirade of unjust criticism he was having to endure and mitigate that insecurity, Adam had to do something. *But what could he do?* If he abandoned the search and flew back on course, he'd gain some momentary relief, but the search would, nevertheless, remain to be undertaken. So all Adam could do was retaliate against and ATTACK the instincts' unjust criticism, attempt to PROVE the instincts' unjust criticism wrong, and try to DENY or block from his mind the instincts' unjust criticism—and he did *all* those things. He became angry towards the criticism. In every way he could he tried to demonstrate his self worth, prove that he is good and not bad—he shook his fist at the heavens in a gesture of defiance of the implication that he is bad. And he tried to block out the criticism—this block-out or denial including having to invent contrived excuses for his instinct-defying behaviour. In short, his ANGRY, EGOCENTRIC and ALIENATED state appeared. Adam's intellect or 'ego' (which is just another word for the intellect since the *Concise Oxford Dictionary* defines **'ego'** as **'the conscious thinking self'**) became 'centred' or focused on the need to justify itself—selfishly preoccupied aggressively competing for opportunities to prove he is good and not bad, to validate his worth, to get a 'win'; to essentially eke out any positive reinforcement that would bring him some relief from criticism and a sense of worth. In summary, Adam unavoidably became SELFISH, AGGRESSIVE and COMPETITIVE. (All the seemingly bad aspects of the intellect and their opposite, seemingly good aspects of the instinct are depicted in the two columns that flank the Adam Stork diagram shown a few pages back.)

But, again, without the knowledge he was seeking, without self-understanding (specifically the understanding of the difference between the gene and nerve-based learning systems that science has given us), Adam Stork had no choice but to <u>resign</u> himself to living a psychologically upset life of anger, egocentricity and alienation as the only three responses available to him to cope with the horror of his situation. It was an extremely unfair and difficult, indeed tragic, position for Adam to find himself in, for we can see that while he was good he appeared to be bad and had to endure the horror of his psychologically distressed, upset condition until he found the *real*—as opposed to the invented or contrived *not*-psychosis-recognising—defence or reason for his 'mistakes'.

Basically, *suffering psychological upset was the price of Adam's heroic search for understanding*. Again, this is the tragic yet inevitable situation any animal would have to endure if it transitioned from an instinct-controlled state to an intellect-controlled state—its instincts would resist the conscious mind's deviating search for knowledge. Adam's uncooperative and divisive competitive aggression—and his selfish, egocentric, self-preoccupied efforts to prove his worth, and his need to deny and evade criticism, essentially embrace a dishonest state—all became an unavoidable part of his personality. Such was Adam Stork's predicament, *and such has been the human condition, for it was within our species that the fully conscious mind emerged.*

And the obvious compensations *we* humans became focused on employing to try to validate ourselves were power, fame, fortune and glory. Yes, we can now understand that materialism—the big house, sparkling chandeliers, fancy clothes, etc—and the financial capital needed to supply it, which Sir Bob protests against, gave humans the fanfare and the glory we knew we were due but which the world in its ignorance of our fundamental goodness would not give us. While spiritual relief (understanding) was still to be found, only material relief was available. Materialism and capitalism became the engine sustaining our necessary but horrifically upsetting search for understanding; ultimately for self-understanding, understanding of our corrupted human condition.

Again, this entire predicament is the fundamental situation *any* animal species would encounter if they developed a fully conscious mind, but in the case of humans we now have to consider how much worse the situation has been for us because *our* instinctive orientation wasn't to a flight path, but to behaving utterly cooperatively and lovingly. In *our* case, when we began experimenting in understanding and were criticised by our instincts and unavoidably responded in an angry, ego-centric and alienated way, we had to endure a further round of criticism, a second hit, a 'double whammy', from our instinctive orientation. Yes, in our necessary search for understanding we were first unjustly condemned for defying our instincts, and then *again* for reacting to that condemnation in a way that was completely counter and offensive to our loving, cooperative instincts. So if Adam Stork had cause to be upset, we had double cause! *No wonder* there has been an underlying volcanic frustration and anger in humans that periodically breaks out in unspeakable acts of violence, cruelty and depravity—and *no wonder* we have had an insatiably greedy appetite for the artificial reinforcement that materialism could give us!!

We can now see how the story of Adam Stork—which describes the primary issue involved in our human condition of the psychologically upsetting battle that emerged between our instincts and our conscious intellect's search for knowledge—has parallels with the pre-scientific Biblical account in the Book of Genesis of Adam and Eve's experiences in the Garden of Eden, except in that presentation when Adam and Eve took the **'fruit' 'from the tree of the knowledge of good and evil'**—went in search of understanding—they were **'banished…from the Garden'** for being **'disobedient'** and becoming 'bad' or **'evil'** or 'sinful' (par. 273). In *this* presentation, however, Adam and Eve are revealed to be the HEROES, NOT THE VILLAINS they have so long been portrayed as. So while humans ARE immensely upset—that is, immensely angry, egocentric and alienated—WE ARE GOOD AND NOT BAD AFTER ALL!!!! (And 'upset' is the right word for our condition because while we are not 'evil' or 'bad', we are definitely psychologically upset from having to participate in humanity's heroic search for knowledge. 'Corrupted' and 'fallen' have been used to describe our condition, but they have negative connotations that we can now appreciate are undeserved.)

We had developed into angry, egocentric and alienated people, but we were never going to accept the implication that we were fundamentally bad, evil, worthless, awful beings; we weren't going to wear that criticism—for if we did, we wouldn't be able to get out of bed each morning and face the world. If we *truly* believed we were fundamentally evil beings, we would shoot ourselves. No, there had to be a greater truth that explained our behaviour and until we found it we couldn't rest. So for our species, it really has been a case of, as the sayings go, **'Give me liberty or give me death'**, **'No retreat, no surrender'**, **'Never back down'**, or, as Sir Bob mentioned, **'death or glory'**. Our conscious thinking self was *never* going to give in to our instinctive self or soul. And so every day as we got out of bed we took on the world of ignorance that was condemning us. We shook our fist at the heavens in defiance of the implication that we are bad. In essence, we said, 'One day, one day, we are going to prove our worth, explain that we are not bad after all, and until that day arrives we are not going to **'back down'**, we are not going to take the ignorant, naive, stupid, unjustified criticism from our instincts. No, we are going to fight back with all our might.' This photograph by Neil Leifer of Muhammad Ali vanquishing Sonny Liston **'is considered to be the greatest sports picture of the 20th century'** (*TIME*, 20 Jun. 2016) precisely because it perfectly expresses every human's desire—destructive as that could be—to vanquish our tormentor, which, at the deepest level, is ignorance of our fundamental worth and goodness.

Photograph by Neil Leifer, 1965, Sports Illustrated

Yes, we were going to fight back against ignorance with *all* our might. And that is what we have done; that is what *every* conscious human who has ever lived has done—and because we did, because we persevered against all that criticism, we have now finally broken through and found the full truth that explains that <u>humans are wonderful beings after all. In fact, not just wonderful but the heroes of the whole story of life on Earth</u>. This is because our fully conscious mind is surely—given its phenomenal ability to understand the world—nature's greatest invention, so for us humans who were given this greatest of all inventions to develop to be made to endure the torture of being unjustly condemned as bad or evil for doing just that, and to have had to endure that torture for *so* long, *some 2 million years* (the time we have likely been fully conscious), *has to* make us the absolute heroes of the story of life on Earth. (pars 65-66 and ch. 3:6 of *FREEDOM*)

Humans were tasked the hardest, toughest of missions, and against all the odds we completed it. Humans *are* the champions of the story of life on Earth. We are *so, so* wonderful! Yes, we can finally understand the absolutely extraordinary paradox that neither Shakespeare nor Pascal could fathom, of how on earth could we humans be **'god'-'like'** in our **'infinite' 'faculty'** of **'reason'** and **'apprehension'**, a **'glor[ious]'**, **'angel'-'like'** **'prodigy'** capable of being a **'judge of all things'** and a **'repository of truth'**, and yet seemingly behave so appallingly that we appear to be **'monster[s]'**, **'imbecile[s]'**, **'a sewer of uncertainty and error'** and **'chaos'**, the **'essence'** of **'dust'**, **'the scum of the universe'**. We have finally made sense of the seemingly nonsensical!!!

And now that we have finally made sense of *the seemingly-impossible-to-explain* paradox of how we humans could be good when all the evidence appeared to unequivocally indicate we were bad, *all* mythology can likewise be made sense of at last. For instance, why was Miguel de Cervantes' 1605 novel *Don Quixote* voted 'The Greatest Book of All Time' by the world's most acclaimed writers in a poll arranged by the Nobel Institute?

Well, *Don Quixote* is the story of an elderly man who gets out of bed, re-names himself 'Don Quixote of la Mancha', dons an old suit of armour, takes up an ancient shield and lance, mounts his skinny old horse, and calls on his loyal but world-weary companion Sancho to join him on the most spectacular of adventures. As I have depicted below, coming across a field of large windmills, the noble knight says, **'Look yonder, friend Sancho, there are…outrageous giants whom I intend to…deprive…of life… and the expiration of that cursed brood will be an acceptable service to Heaven'**. And it is in that vein that their crazed and hopeless adventure went on—gloriously doomed battle after gloriously doomed battle; feeble beings charging at and trying to vanquish the **'outrageous giant'** ignorance-of-the-fact-of-our-species'- fundamental-goodness! <u>But that has been the lot of every human for some 2 million years!</u> Wave after wave of *quixotic* humans have thrown themselves at that **'outrageous giant'** of ignorance for eons and eons, as bit by tiny bit they accumulated the knowledge that would finally make the redeeming explanation of our human condition possible! (par. 67)

Drawing by Jeremy Griffith © 2015 Fedmex Pty Ltd

The following painting by J.M.W. Turner is another powerful portrayal of how absolutely incredibly HEROIC the human race has been, huddled together for some reassurance, and with few provisions, while we struggled as best we could through *2 million years* of terrifying darkness and tumultuous storms to acquire that relieving knowledge. To be given a fully conscious brain, the marvellous computer we have on our heads, but *not* be given the program for it and instead be left '**a restless wanderer on the earth**' (Bible, Gen. 4:14) searching for that program/understanding in a dreadful darkness of confusion and bewilderment, most especially about our worthiness or otherwise as a species, *was* the most diabolical of tortures. As the great denial-free thinking prophet Isaiah put it: '**justice is far from us, and righteousness does not reach us. We look for light, but all is darkness; for brightness, but we walk in deep shadows. Like the blind we grope along the wall, feeling our way like men without eyes…Truth is nowhere to be found**' (Bible, Isa. 59).

J.M.W. Turner's *Fishermen at Sea*, 1796

Turning to more recent cultural depictions, Turner's *Fishermen at Sea* recalls the lyrics that *so* plead the terrible agony of our species' seemingly lost, lonely and meaningless condition—**'How does it feel to be on your own, with no direction home, like a complete unknown'**. The profundity of these words have understandably led to the 1965 song from which they come, *Like a Rolling Stone* by that prophet of our time, Bob Dylan, to be voted in 2011 as the greatest of all time by that arbiter of popular music, *Rolling Stone* magazine. (par. 278)

Yes, how *incredibly* heroic have we humans been—and how wonderful, beyond-the-powers-of-description, is it to now have freed ourselves from that horrific situation where **'We look for light, but all is darkness...Truth is nowhere to be found'**!

In summary, we humans had *no* choice but to persevere with our search for knowledge and suffer the psychologically upset state of being angry, alienated and egocentric until we could develop the scientific method and through that vehicle for enquiry find the redeeming explanation for our species' upset condition of the difference between the gene and nerve-based learning systems—the key insight that reveals we humans are good and not bad after all. Yes, until science made it possible to explain the difference between the gene-based natural selection process that gives species orientations and the nerve-based conscious mind that needs to understand, every time we tried to think about our corrupted condition the only conclusion we could come to was that we conscious humans had wilfully destroyed paradise. So science is the liberator—the messiah—of humanity from the agony of the human condition.

And importantly, now that we *are* able to understand from scientific first principles that upset is not an 'evil', worthless, bad state, but an immensely heroic state, we can know that while, inevitably, all humans are variously upset from their different encounters with, and degrees of engagement in, humanity's epic search to find knowledge, ALL HUMANS ARE

EQUALLY GOOD. Everyone has variously had our species' original innocent instinctive state corrupted, everyone is variously angry, egocentric and alienated, but everyone is good, and not just good but a hero of the story of life on Earth! No longer does humanity have to rely on dogmatic assertions that **'all men are created equal'** purely on the basis that it is a **'self-evident'** truth, as the United States' Declaration of Independence asserts, because we can now *explain, understand* and *know* that the equality of goodness of all humans is a fundamental truth. We can now understand *why* each individual, gender, age, race, country, civilisation and culture is equally good and worthy, and that no one is superior or inferior, and that *everyone* deserves the **'rights'** of **'life, liberty and the pursuit of happiness'**. Indeed, through this understanding, the whole concept of good and bad, basically of guilt, disappears from our conceptualisation of ourselves. Compassion, love and understanding for our human situation has finally been achieved— which is what Sir Laurens van der Post recognised was needed when he variously wrote in his books: **'True love is love of the difficult and unlovable'**; and, **'how can there ever be any real beginning without forgiveness?'**; and that **'Only by understanding how we were all a part of the same contemporary pattern** [of wars, cruelty, greed and indifference] **could we defeat those dark forces with a true understanding of their nature and origin'**; and that **'Compassion leaves an indelible blueprint of the recognition that life so sorely needs between one individual and another; one nation and another; one culture and another. It is also valid for the road which our spirit should be building now for crossing the historical abyss that still separates us from a truly contemporary vision of life, and the increase of life and meaning that awaits us in the future.'** Yes, one day, which has now arrived, there had to be, to quote The Rolling Stones' lyrics, **'Sympathy for the devil'**; one day, we had to find the reconciling, compassionate, healing understanding of the dark side of human nature. One day, *'The Marriage of Heaven and Hell'*, as William Blake titled his famous book, had to occur. (par. 290)

So we now *know* that there was a *good* reason why we humans became corrupted—we *had to* search for knowledge, ultimately sufficient knowledge to explain the human condition. And it is necessary to emphasise here that there was no real way out of our psychologically upset condition until we found that redeeming understanding. Searching for knowledge made us angry, egocentric and alienated, but that was the price we had to be prepared to pay. At any time we could stop our upsetting search for knowledge—fly back on course in the Adam Stork analogy—which would make us feel good, stop the criticism from our instincts, but that was an abandonment of our fundamental responsibility as conscious beings to keep persevering with our search for knowledge and suffer all the upset that resulted from doing that until we found the knowledge we were after of the explanation of the human condition.

This is a key point to make, because it allows us to understand the real merits and liabilities of the left and right wings of politics. It turns out that of the twin political problems we have, of the brutality of the 'need-to-continue-the-corrupting-search-for-knowledge-free-from-condemning-idealism' right-wing, and the dishonest and deluded 'I'm-being-good-by-flying-back-on-course' pseudo idealism of the left-wing, it was actually the corrupting individualistic, materialistic and capitalistic right-wing that held the moral high ground, that was doing the right thing, being *genuinely* idealistic, because it was persevering with the necessary search for knowledge, whereas the pseudo idealistic dogma of the left has actually been subverting the human journey (ch. 3:9)! Such is the great paradox of the human condition that we can now finally understand: the truth was not as it appeared—the corrupting journey Adam Stork/the right-wing were championing made them the heroes not the villains! As it says about the great paradox of the human condition in the musical *Man of La Mancha* that is based on the story of Don Quixote, we humans had to be prepared to **'march into hell for a heavenly cause'** (par. 68).

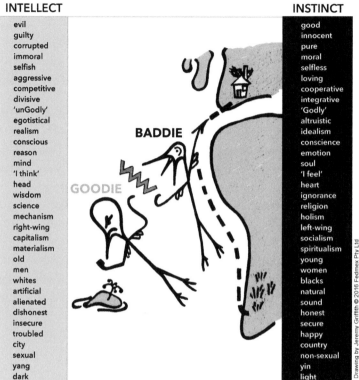

INTELLECT	INSTINCT
evil	good
guilty	innocent
corrupted	pure
immoral	moral
selfish	selfless
aggressive	loving
competitive	cooperative
divisive	integrative
'unGodly'	'Godly'
egotistical	altruistic
realism	idealism
conscious	conscience
reason	emotion
mind	soul
'I think'	'I feel'
head	heart
wisdom	ignorance
science	religion
mechanism	holism
right-wing	left-wing
capitalism	socialism
materialism	spiritualism
old	young
men	women
whites	blacks
artificial	natural
alienated	sound
dishonest	honest
insecure	secure
troubled	happy
city	country
sexual	non-sexual
yang	yin
dark	light

BADDIE

GOODIE

Drawing by Jeremy Griffith © 2016 Fedmex Pty Ltd

Understanding the human condition means the tables are turned on our perception of reality. The self-adjusting conscious mind, all the manifestations of which are shown in the left-hand column of the Adam Stork diagram above, turns out to be the hero not the villain of the story of life on Earth, the 'goody' — while our unjustly condemning instinctive soul, and all its manifestations (as shown in the right-hand column of the Adam Stork diagram), turns out to have been the cause of all our upset anger, egocentricity and alienation, the 'baddy'.

Again, with the reconciling understanding of the human condition now found, *neither* aspect of our divided self is good or bad. Our instinct and intellect are reconciled. For example, as explained in the aforementioned chapter 3.9 of *FREEDOM*, 'The end of politics', the very basis of the 'twin political problems' of the left-wing and right-wing in politics is now completely obsoleted—understanding brings the whole ugly, intransigent business of politics to an end.

It needs to be emphasised here that when the blinds are finally drawn on the human condition the light inevitably reveals a universe of previously denied truths—a litany of heresies. In fact, the arrival of understanding of the human condition brings about honesty day or truth day or exposure day or transparency day or revelation day, or so-called judgment day—but as I have explained, it is a day of compassionate understanding, not of cruel judgment. (ch. 9:3)

In addition to the previously not properly understood and therefore unappreciated role of the right-wing in politics, and the exposure of the left-wing as fundamentally irresponsible and dishonestly deluded about being ideally behaved, the following are some of the heresies (which are fully explained in *FREEDOM*) that we now have to adjust to:

- that the human race has become an extremely upset, alienated wreck (see for example pars 123 and 182);

- that dishonest, human-condition-avoiding mechanistic science has been leading the human race to a state of terminal alienation (ch. 2:4);

- that humans have a cooperative, loving instinctive heritage, not a brutish, savage and aggressive one (ch. 5);

- that nurturing was the prime-mover in human development, and is still critically important (ch. 5);

- that sex as humans now practise it is not about trying to reproduce our genes but rather it is a psychologically upset behaviour where we resentfully attack innocence for its seeming criticism of our lack of innocence (ch. 8:11B);

- that there are differences between 'races' (ethnic groups) in their levels of corruption of, and alienation from, our species' original innocent instinctive state. Some 'races' are less innocent than others; or to say it another way, some 'races' are more adapted to the upset state of the human condition. This is an inevitable result of the upset state of the human condition having emerged in our species; some 'races', just as some individuals, will have been more exposed to upset and become more adapted to it. Everyone is variously upset. Everyone has variously lost their innocence. As soon as you have a struggle of any sort, those involved are all going to be variously exposed to it and thus variously affected by it. It can't be any other way. But with the human condition at last resolved all the upset in humans can now be repaired. The human race is now redeemed and everyone can be rehabilitated (ch. 8:16E);

- that women are not 'mainframed' to the battle of the human condition in the same way men are; they are not as understanding of and thus appreciative of the battle as men are, simply because with women naturally occupied with the critical task of nurturing, the horrifically upsetting battle of searching for knowledge had especially to be taken up by men, and, with that battle now finally won, what is revealed is that men particularly are the heroes of the story of life on Earth. Men are now explained and redeemed, and, more importantly, the historic so-called 'battle of the sexes', the lack of understanding between men and women, is brought to an end (ch. 8:11B);

- And (as is explained in chapter 4) that God is the teleological, order-developing, physical law of negative entropy, and (as explained in paragraphs 877 and 929-935) that Christ was not a deity but a human like everyone else, only sufficiently loved in his infancy to not have resigned to a life of dishonest denial of the issue of the human condition. So religions are now obsoleted with understanding, and, as will be described next in Part 3 of this book, replaced with an infinitely more powerful and effective way of living a meaningful, truth-aligned life, free of your upset.

So again, the truth is not as it appeared; the previously condemned intellect and all its manifestations turn out to be good and not bad, while the instinctive state and all its manifestations turn out to have been unjustly condemning of humanity's heroic search for knowledge. This 'turning of the tables' will be an immense shock; as another truthful-talking prophet of our time, Bono of the band U2, said recently, **'truthfulness will blow things apart'**. And yes, we will initially find all that exposure of our cave-living evasions, denials, lies and delusions difficult to cope with; as George Bernard Shaw said, **'All great truths begin as blasphemies'** (par. 591). However, this 'blasphemous' stage, when all the previously denied truths are suddenly revealed and we are in shock, won't last long once the wonderfully liberating potential of these understandings that is described in chapter 9 and summarised next in Part 3 catches on.

And *again*, with the reconciling understanding of the human condition now found, neither aspect of our divided self is good or bad. Our instinct and intellect *are* reconciled. For example, as already explained, the very basis of the left-wing and right-wing in politics is now obsoleted, as is the prejudiced view of some individuals or 'races' or genders or cultures being inferior or superior to others—understanding heals the human race.

Yes, our ability now to explain and understand that we

humans are actually good and not bad enables *all* the upset and associated misunderstanding and dysfunction that resulted from being unable to explain the source of our divisive condition to subside and disappear. Finding understanding of the human condition is what rehabilitates and transforms the human race from its psychologically upset state. In fact, the word **'psychiatry'** literally means **'soul-healing'** (derived as it is from *psyche* meaning 'soul' and *iatreia*, which means **'healing'**) — but we have never before been able to 'heal our soul', to truthfully explain to our original instinctive self or soul that our fully conscious, thinking self is good and not bad and, by so doing, reconcile and heal our split selves; **'get our old natural selves to join with our other conscious, wilful, rational, scientific selves'**, as Sir Laurens van der Post said we had to do. As Professor Harry Prosen says in his Introduction to *FREEDOM* about the psychological effect of this human-race-saving and thus world-saving reconciling under-standing of ourselves: **'I have no doubt this biological explanation of the human condition is the holy grail of insight we have sought for the psychological rehabilitation of the human race.'** It brings about the **'wholeness for humans'** that the great psychoanalyst Carl Jung was forever pointing out **'depends on the ability to own our own shadow'**, to understand the dark side of ourselves (par. 72). As depicted in the final Adam Stork diagram overleaf, it reconciles our intellect and instinct and all their manifestations, thus enabling humans to become sound, secure and **'whole'** again. It's been an astounding journey that humanity has negotiated from happy innocent ignorance, to a state of corrupted upset, to happy upset-free soundness — a round of departure and return that was perfectly anticipated by T.S. Eliot when he wrote that **'We shall not cease from exploration and the end of all our exploring will be to arrive where we started and know the place for the first time'** (par. 307).

INTELLECT		INSTINCT
evil		good
guilty		innocent
corrupted		pure
immoral		moral
selfish		selfless
aggressive		loving
competitive		cooperative
divisive		integrative
'unGodly'	**RECONCILED AND**	'Godly'
egotistical	**TRANSFORMED**	altruistic
realism		idealism
conscious		conscience
reason		emotion
mind		soul
'I think'		'I feel'
head		heart
wisdom		ignorance
science		religion
mechanism		holism
right-wing		left-wing
capitalism		socialism
materialism		spiritualism
old		young
men		women
whites		blacks
artificial		natural
alienated		sound
dishonest		honest
insecure		secure
troubled		happy
city		country
sexual		non-sexual
yang		yin
dark		light

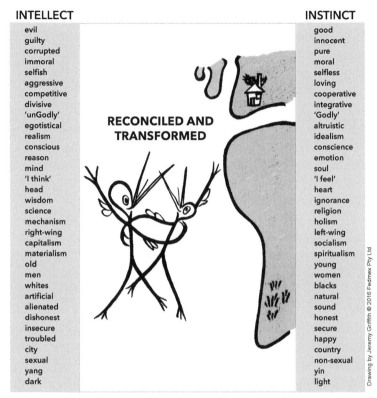

Drawing by Jeremy Griffith © 2016 Fedmex Pty Ltd

While it will take a number of generations for all our species' upset psychosis to be completely healed, <u>the most important point to take on-board is that with this clarifying understanding of our fundamental goodness, all of our psychologically defensive angry, alienated and egocentric behaviour is immediately made redundant. As I explain in chapter 9 of *FREEDOM*, and summarise next in Part 3 of this book, with the great burden of guilt lifted from the human race, the old insecure, defensive and retaliatory upset life that went with it is finally over, which means we can all immediately leave that way of behaving behind as finished with; we can *immediately*</u>

change from living competitively to living cooperatively—which is the great transformation in behaviour that Sir Bob has long dreamed of.

As Sir Bob's fellow rock star, and fellow prophet of our time, Bono anticipated in his songs, everyone will now be able **'to feel sunlight** [of liberating understanding] **on my face, see that dust cloud** [of all our upset behaviour] **disappear without a trace'**. We will be able **'to take shelter from the poison rain...high on a desert plain** [in a wonderful ego-less new world] **where the streets have no name'**; a place where **'there will be no toil or sorrow...no time of pain'** (par. 218), and everyone will sing **'I've conquered my past, the future is here at last, I stand at the entrance to a new world I can see. The ruins to the right of me, will soon have lost sight of me. Love** [in its purest form, which is truth, has] **rescue**[d] **me'** (par. 1288).

Indeed, when the singer Jim Morrison wrote and sang **'At first flash of Eden we race down to the sea. Standing there on freedom's shore, waiting for the sun...waiting...to tell me what went wrong'** (par. 218), he was looking forward to the arrival of 'FREEDOM: The End Of The Human Condition' with its beautiful cover showing the rising sun and people jumping for joy in front of it.

Yes, now we really *can* have the dancing in the street that Sir Bob didn't believe was possible when he was using the now obsoleted, dishonest biological way of thinking about human behaviour!

I concluded my presentation at the Royal Geographical Society with the playing of Martha & the Vandellas' version of *Dancing in the Street*, the lyrics of which also anticipate this most wonderful, wonderful of breakthroughs that science has finally made possible: **'Calling out around the world, are you ready for a brand new beat? Summer's here and the time is right for dancing in the street.'** (Note, many more exciting anticipations from songs and literature of the arrival of understanding of the human condition and the liberation of the human race that it brings about can be read in pars 1283 to 1292 of *FREEDOM*.)

As I mentioned at the beginning of Part 1, on my return to Australia following the launch of *FREEDOM* at the Royal Geographical Society, I gave an additional presentation explaining how this new truthful biology brings about the <u>immediate</u> transformation of the human race from the present tortured, insecure, human-condition-stricken existence to a fabulous, secure, human-condition-free life—thus 'stopping the madness, filling the void, bringing dignity to humans, giving the world hope, ending the despair and nourishing our dreams' that Sir Bob so yearned for. The expanded transcript of this presentation follows as Part 3 of this book.

Part 3

The Resulting Immediate Transformation Of The Human Race

(Note, Part 3 is a condensation of chapter 9 of *FREEDOM*.)

So, how is this fully accountable, true explanation of the human condition able to *immediately* transform your, and everyone else's, life? How does the arrival of the **'summer'** of self-understanding make it **'the time' 'for a brand new beat'** and **'dancing in the street'**? How does the arrival of **'the sun'** on **'freedom's shore'** that we have been **'waiting'** for, to explain the human condition and **'tell me [us] what went wrong'**, bring about the **'first flash of Eden'**? Essentially, how does this explanation take us all back home again to our species' original Edenic state of togetherness and happiness, only this time with the ability to understand that state or, as T.S. Eliot wrote, **'know the place for the first time'**?

The answer is a natural and logical consequence of our ability to finally be able to explain and understand the whole journey humanity has been on and our situation now that understanding of the human condition has been found. To summarise, it was through the process of nurturing that our distant primate ancestors were able to develop a completely cooperative and loving existence, like the bonobos are in the process of developing today. We then became conscious, at which point a battle for the management of our lives began between our self-adjusting conscious mind and our already established instinctive orientations. Unable to explain and understand *why* we had to challenge our instincts, we then became psychologically retaliatory, defensive and insecure—upset sufferers of the angry, egocentric and alienated human condition. But now, after 2 million years of searching, we have finally found the explanation and understanding for *why* we had to challenge our instincts, which ends the need for humans to be retaliatory, defensive and insecure. The **'sun'**-light of understanding brings an end to our psychologically upset angry, egocentric and alienated existence. The battle to establish our goodness and worth is over, which means we can finally 'put the sword down'; the game is won so we can all 'head for the showers', or however else we like to describe our own and our species' emancipation. A horrifically upset and stressful way of living ends and a new life free of all that upset begins. Humans can finally return from a tormented psychotic existence to health and happiness.

The healing, however, of our species' upset, psychotic state and the complete return to health and happiness for the human race that this breakthrough makes possible will, as has already been mentioned, take a number of generations. This is because psychoses usually take years to heal with understanding, so it follows that in the case of the *largest* psychosis of all of the human condition, that healing process will naturally be a generational process. The big question that remains then, is how is this arrival of understanding of the human condition able to *immediately* transform your and every other humans' life? How do you immediately transform your life from an insecure, psychologically embattled, upset state where you were deriving reinforcement for your sense of goodness and worth by winning as much power, fame, fortune and glory as you could, or by taking up support of some pseudo idealistic cause like religion or socialism or environmentalism or feminism or political correctism or post-modernism or multiculturalism or anti-capitalism?

To address this, I need to really drive home the fundamental situation you and every other human is now in: the battle for every human to prove his or her worth is *over*. In fact, to continue fighting to prove we are good and not bad when our fundamental goodness has been established is not only obviously pointless but also unnecessarily destructive of ourselves, everyone around us and of our world. That way of living is completely obsoleted, finished with. But when you are already resigned to living in denial of the human condition and have spent your entire adult life trying to sustain your sense of worth by seeking power, fame, fortune and glory, or by supporting upset-relieving pseudo idealistic causes, *can* that way of living suddenly be abandoned? *Can* humans give up a highly habituated existence from living competitively and selfishly to living cooperatively and lovingly, or, in the case of supporters of pseudo idealism, change from masquerading as an idealist to living honestly with the truth of their corrupted condition?

The answer is a resounding yes: humans *can* instantly change from living one way to living a completely different way, and for evidence that you and every other human can make such a change, we need only look at what occurs in religious conversions (although as will be made very clear, the change this explanation makes possible is *very* different to a religious conversion). For example, in the case of a conversion to Christianity, when you drive past a Christian church and see one of the signs they place out the front with 'Jesus is the answer', or 'Be born again through Christ', or 'Christ is our Saviour', or 'Give your life to Christ and be redeemed', we are all well aware that what's on offer is the option for people to give up their personal struggle with the human condition and take up a life of relieving deferment to Christ. While we may not have experienced a religious conversion ourselves, most of us will know someone who has. For example, *The Simpsons* cartoon series, a show that is immensely popular because it incisively reflects life's realities, features the character Ned Flanders, a Christian who has transcended his human-condition-afflicted, upset existence and been 'born-again' (as the transformation is revealingly termed) to living cooperatively and lovingly again by deferring to, and living through, Christ. Ned's neighbour Homer Simpson, on the other hand, is still living out a human-condition-embattled, upset life to the full, drinking and brawling and crashing about in hapless efforts to validate himself. In one episode, for example, Ned lends Homer his lawnmower, which Homer ends up wrecking without remorse. But rather than getting angry or defensive, Ned simply accepts Homer's behaviour—he is the 'goody-goody' Christian while Homer is a prime example of an upset human, living out the battle of the human condition to the full: he is massively angry, egocentric and alienated.

So we humans *do* have the ability, if we so choose, to transcend our struggling-to-maintain-our-sense-of-self-worth, human-condition-stricken existence and live through the support of a set of ideal values and practise more loving behaviour than our upset self wants to practise. To use the Adam Stork analogy, Adam could always fly back on course, give up his upsetting pursuit of knowledge and simply obey his instinctive orientations, which, for us humans, is to be cooperative and loving—the relatively innocent, uncorrupted and upset-free way of living that is embodied in the lives of the great prophets around whom religions are founded.

Humans *can immediately* abandon a must-win, competitive, selfish and aggressive power, fame, fortune and glory-seeking way of living and instead live in a way that is supportive of, and consistent with, a cooperative, selfless and loving way of living. However, what needs to be explained and emphasised straight away is that while the transcendence of your psychologically embattled, upset self that is involved in the new human-condition-resolved, upset-behaviour-obsoleted, cooperative, selfless and loving conversion also occurs in a religious conversion, in every other respect it is *very* different.

FIRSTLY, and most significantly, this human-condition-resolved, upset-behaviour-obsoleted new way of living is *not* a fundamentally irresponsible and weak abandonment of the human race's battle to find understanding, as was the case when deferring to the sound life of a prophet in religion, or, for that matter, to taking up support of any other form of pseudo idealism, such as socialism or environmentalism or feminism or political correctism or post-modernism or multiculturalism or anti-capitalism. No, in stark contrast to these ways of *coping* with the human condition, this human-condition-resolved, upset-behaviour-obsoleted new way of living is *a completely responsible and entirely strong way of living*.

As has already been explained, the great problem with obeying our species' instinctive orientation to behave cooperatively and lovingly (to 'flying back on course' in the Adam Stork analogy) is that up until now it meant you had personally given up participating in humanity's corrupting but heroic search for knowledge, ultimately for self-knowledge, the understanding of the human condition that alone could free us from that condition. While it was necessary for people to abandon the corrupting search for knowledge when their experiences from and/or participation in that great battle had resulted in them having become too corrupt and destructive, in terms of persevering towards our species' ultimate goal, it *was* fundamentally an irresponsible act of weakness—as the philosopher Friedrich Nietzsche recognised, **'There have always been many sickly people'** who **'have a raging hate for the enlightened man and for that youngest of virtues which is called honesty…You are not yet free, you still *search* for freedom. Your search has fatigued you… But, by my love and hope I entreat you: do not reject the hero in your soul! Keep holy your highest hope!'** (par. 302 of *FREEDOM*). However, now that our **'highest hope'** of finding understanding of the human condition has been realised, it is no longer an act of weakness to abandon the battle; in fact, it's now not only an act of strength, it is the *only* responsible strategy for a human to adopt. Everyone *should* abandon the battle of the human condition because it is now won and over. There is *no reason at all* now not to give up the old 'must-prove-your-worth', power, fame, fortune and glory way of living and participate in this completely responsible and non-pseudo-idealistic new way of living that solves all our and the world's problems.

If we look at the Ned Flanders/Homer Simpson situation again, as a 'born-again' Christian Ned is typically portrayed as having a self-satisfied, 'I-occupy-the-moral-high-ground' attitude over the still-human-condition-embroiled Homer Simpson. Ned's smug posturing drives Homer crazy with frustration because Homer intuitively knows Ned is deluding himself in thinking his Christianity gives him the moral high ground—that he is the more 'together', sound, better person and is on the right track—but without understanding of the human condition Homer couldn't explain *why* Ned is so extremely deluded and dishonest in his view of himself. Homer couldn't explain and thus reveal the truth that *real* idealism and the truly on-track, moral high ground lay with continuing the upsetting battle to find knowledge, and that Ned had become so upset, so *un*-sound, that he had to abandon that all-important battle and leave it to others to continue to fight, including Homer. Worse, in abandoning the battle and, for example, deludedly maintaining that doing so was the solution to the world's problems, Ned has effectively sided against those still trying to win the battle, adding substantially to the opposition they had to overcome. If Homer could have explained the situation, he would have said to Ned: 'Listen Ned, you love Christ and he loves you, and you're a goody-goody, and I'm one upset, corrupted, dysfunctional dude, but Ned, I'm still out there participating in humanity's heroic battle to find knowledge, so I'm a bold and brave legend and you're a worn-out quitter.'

I should explain that while religion is a form of pseudo idealism because in adopting it you're *fundamentally* abandoning the upsetting battle to find understanding, of all the different forms of pseudo idealism, religions have, by far, been the *least* destructively dishonest and deluded because of the honesty of the prophets' words about the corrupted state of the human condition. In fact, as the human race has become extremely upset, religions have been losing favour in the West because that honesty, and the emphasis on guilt that results from it, has become too unbearably confronting and condemning; the great value of religion has become its liability. Chapters 8:16H to 8:16Q of *FREEDOM* document how as upset has increased there has been a progression to ever more truth/guilt-stripped forms of pseudo idealism that humans have found to adopt—to simplistic, fundamentalist expressions of religion, to a sequence of non-religious pseudo idealistic causes; in particular from socialism to the New Age movement to the feminist movement to the environment movement to the politically correct movement and then to the completely truth-stripped post-modern, deconstructionist movement. Shortly, in the 'Sixth difference between this new transformed way of living and religion', it will be explained that the almost complete removal of dishonesty and delusion involved in this new transformed way of living is one of its greatest benefits; with upset defended, the need for the dishonesty and delusion associated with living in denial of the human condition is almost completely removed. The human race is finally able to leave Plato's horribly soul-destroying/ psychosis-producing/alienating dark cave of dishonest denial. (I explain much more about the critically important contribution religion has made to the human journey in chapters 8:15 and 8:16 of *FREEDOM*.)

So, the situation now with the new human-condition-resolved, upset-obsoleted new way of living is that it is *not* an act of irresponsibility or weakness to abandon the battle, because that great battle has been won, understanding of the human condition has been found, so to continue that now-obsolete battle would be illogical and thus irresponsible and thus the weak thing to do. If we imagine then that Ned has taken up the human-condition-resolved, upset-obsoleted, transformed new way of living, Homer would have no grounds to criticise him. In this scenario, the tables are turned: Ned would be in the position to censure Homer.

We are rational creatures, and so when all the logic points to there being only one response we can make, namely the human-condition-resolved, upset-obsoleted, transformed new way of living, then that is the response the human race has to take. No revolution dogmatically, and even physically, imposed upon us by others, as has pretty much been the case with revolutions in the past, just a peaceful revolution through the power of logic, knowledge, understanding. Of course, while it is the power of the logic, the rationale, that makes the new transformed way of living irresistible, the adoption of this way of living is also made irresistible by the absolutely wonderful transforming effect it has on people. Indeed, before long those still living in the old embattled, have-to-prove-your-worth, or deluded, do-good-to-feel-good pseudo idealistic ways of living will feel like they have been caught wearing last year's fashions! There is such a stark difference between the free, authentic, expansive, enthralled-with-life, transformed existence and the old embattled existence, where your mind is narrowly focused and preoccupied working all the angles every second of every day, defensively seeking validation and reinforcement, that those living in the old ways will seem like a different species. Compared to the new expansive, all-exciting, free, transformed state, those still stuck in the old embattled paradigm *will* be like sad, forgotten ghosts trapped in an obsoleted state.

When the scientist-philosopher Pierre Teilhard de Chardin wrote, '**The Truth has to only appear once...for it to be impossible for anything ever to prevent it from spreading universally and setting everything ablaze**' (par. 1236 of *FREEDOM*), he was anticipating the arrival of the all-transforming and all-exciting understanding of the human condition. I want you to go outside and look up towards the horizon because, I kid you not, what you're soon going to see appear, from one end of that horizon to the other, is an army in its millions coming to do battle with human suffering and its weapon will be understanding! With the world's preoccupation with finding self-understanding now resolved, and all the associated need for self-aggrandisement satiated, our lives and the world's resources will be freed to finally look after each other and our world. So while you might have seen some wonderful things in your life, like fields of flowers or gorgeous sunrises or music festivals of people dancing, you won't have seen anything even remotely as incredible, wonderful and joyous as the human race coming home together to peace and happiness! And it has to be remembered that behaving cooperatively and lovingly is our original natural way of behaving, so when we all realise it is now the only legitimate and responsible way to behave, then that way of behaving will come thundering through. Take the 'boot' of repression off the 'throat' of our soul and it will 'breathe' freely once more! Again, when that truth-talking prophet of our time, Bono, wrote about and sang of a time when we would be able '**to feel sunlight** [of liberating understanding] **on my face, see that dust cloud** [of all our upset behaviour] **disappear without a trace**', a time when we will be able '**to take shelter from the poison rain...high on a desert plain** [in a wonderful ego-less new world] **where the streets** have no name'; a place where **'there will be no toil or sorrow...no time of pain'**, and everyone will sing **'I've conquered my past, the future is here at last. I stand at the entrance to a new world I can see. The ruins to the right of me, will soon have lost sight of me'**, he was anticipating the arrival *at last* of the human-condition-free world!

The **SECOND** fundamental difference between this new transformed way of living and religion is that while religions were about supporting the *embodiment* of the ideals in the form of the prophet/s around which they were founded, this new way of living is about supporting the *understanding* of the ideals and why we fully conscious humans have historically not been able to live by them. <u>This is the end of dogma, faith and belief, and the beginning of understanding and knowing</u>.

In his acclaimed 1969 BBC television documentary series *Civilisation*, the eminent historian Kenneth Clark mentioned that **'People who hold forth about the modern world often say that what we need is a new religion. It may be true but it isn't easy to establish.'** Saying **'that what we need is a new religion'** is really an acknowledgment of the fundamental need, if we are to truly solve the world's problems, for humans to change from living competitively, aggressively and selfishly to living cooperatively, lovingly and selflessly. And, saying **'but it isn't easy to establish'** **'a new religion'** is really, deep down, a recognition that what is needed for this great change to occur is for understanding of the human condition to be found, because while it wasn't found the upsetting battle to find it had to continue. Humans have always intuitively known they had to suffer living an upset, corrupted existence until we found self-understanding—we have intuitively been aware that, as it says in the *Man of La Mancha*, we had to **'march into hell for a heavenly cause'**. When it was reported that **'The** [Tibetan Buddhist leader, the] **Dalai Lama believes secular ethics, not religion, is best placed to assist the "moral crisis" facing the world's people...**[because secular ethics] **respected all traditional faiths as well as non-believers'**, and quoted the Dalai Lama saying that **'"Some people—some my** friends—believe moral ethics must be based on religious faith…[But] No matter how wonderful a religion, (it will) never be universal. The crisis is universal—now the remedy must also be universal'", the Dalai Lama was also, at base, recognising the need for the reconciling understanding of the human condition to arrive because *only* that could supply the 'universal[ly]' acceptable 'remedy' of 'secular ethics' he was acknowledging the world needs. (par. 1222)

When John Lennon (who I consider to be another truth-saying prophet of our time, like Bob Dylan, Bono and Sir Bob Geldof) anticipated in his 1971 song *Imagine* the arrival of a time when the human condition is resolved and 'the world will be as one', he too was envisioning a world free of the condemning differentiation of good and evil, a time when there would be 'no heaven [above us and] no hell below us', a world liberated from the insecurity of the human condition and thus the need for religion, where, as he sang, there will be 'Nothing to kill or die for, and no religion too…all the people living life in peace…No need for greed or hunger, a brotherhood of man…all the people sharing all the world'.

This second differentiation to religion leads directly to the **THIRD** immense difference between this new transformed way of living and a religion, which is that <u>unlike religion, there is no involvement or emphasis on guilt, because guilt—and the whole notion of 'good and evil'—has been eliminated forever with the reconciling understanding of the human condition</u>.

And it raises the **FOURTH** fundamental difference about this new transformed way of living, which is that <u>there is no deity involved, no emphasis on 'God', or deference to any one personality. In fact, there is no worship of any kind</u>.

The **FIFTH** difference this new transformed way of living has to religion also results from it being based on knowledge, not faith or dogma, which is that <u>dogma can't heal upset, only understanding can do that</u>. Ultimately de-braining ourselves wasn't going to work. We needed brain food not brain anaesthetic. We needed confusion-relieving understanding. *We needed answers.*

Yes, in complete contrast to religion, the new transformed way of living is concerned with what happens *after* the liberating understanding of the human condition is found, which is the advancement of the human race from the bewildered, insecure, human-condition-stricken state of mind to an existence free of that deadening pain in the brain.

In fact, it is impossible to express just *how* redeeming, empowering and thus rehabilitating it is for the human race to now have a bedrock of first principle scientific knowledge that allows us to understand our fundamental goodness. For humans, knowledge is the ultimate form of power, and this knowledge we now have has reached right to the 'bottom of the well' of who we are, and from there, brought an end to the underlying insecurity, confusion, frustration, worry, doubt, shame, guilt, anxiety and angst that has plagued human life. It has, in short, opened up a whole new world free of the human condition. Yes, the human-condition-understood, upset-obsoleted, new transformed way of living, and the psychological healing of the human race that it allows to occur over a few generations, *completely* changes the human race from a state of troubled upset to a state of secure soundness. In fact, it *metamorphoses*—it matures—the whole human race from insecure adolescence to secure adulthood!

The **SIXTH** very important difference between the new transformed way of living and religion applies both to the power, fame, fortune and glory obsoleted way of validating ourselves and to religion and the various other pseudo idealistic ways of relieving ourselves of our upset. This key difference is that the new transformed way of living is virtually free of the human-condition-avoiding dishonesty and delusion involved in those ways of living. With our upset lives now defended, we not only don't have to prove our worth, we don't have to deny any truths about our immensely corrupted condition—and as was outlined in the bullet points at the end of Part 2, there have been *a great many* truths that resigned humans have been living in denial of. For example, as explained in Part 1, science is supposed to be our vehicle for the pursuit of knowledge, but it is now so saturated with lies that it has become a **'citadel...in ruins'** (as Koestler described it).

As I emphasised at the end of Part 2, when the blinds are finally drawn on the human condition the light inevitably reveals a universe of previously denied truths, a litany of heresies. All the lies and delusions we humans have been using to protect ourselves from exposure of our corrupted condition while we couldn't truthfully explain that condition are suddenly revealed. The truth reveals the lies, as it must. In fact, as I mentioned there, the arrival of understanding of the human condition brings about honesty day or truth day or exposure day or transparency day or revelation day, or the so-called 'judgment day' that is referred to many times in the Bible. While, as I stressed, this all-exposing 'judgment day' is a time of compassionate understanding not condemnation, it naturally does come as a great shock to have the extent of our own and our species' now *extremely* psychologically upset angry, egocentric and alienated condition suddenly revealed.

This paradox of being wonderfully liberated but at the same time suddenly frighteningly exposed was recognised by the prophet Isaiah when he said that the liberation that **'gives you relief from suffering and turmoil and cruel bondage...will come with vengeance; with divine retribution...to save you. Then will the eyes of the blind be opened and the ears of the deaf unstopped...Your nakedness will be exposed and your shame uncovered...on the day of reckoning'**. In the Koran, the prophet Muhammad also referred to **'the Day of Reckoning'** and **'the Last Judgement'**, similarly stating that **'on that day, the Terror shall come to pass, and heaven shall be split...On that day you shall be exposed, not one secret of yours concealed.'** (par. 1153)

Humans typically need time to adjust to change, and the arrival of understanding of our corrupted condition is such an immensely significant and sudden event that our ability to adjust to it can't but be overwhelmed. In his famous 1970 book, *Future Shock*, Alvin Toffler was remarkably prescient in his anticipation of this time when understanding of the human condition would emerge and humans would suddenly be faced with (as he put it) **'the shattering stress and disorientation that we induce in individuals by subjecting them to too much change in too short a time'** (par. 1152).

George Bernard Shaw was another who succinctly described this problem of the sudden change that the arrival of a great insight brings and the inevitable difficulty, and natural resistance, people have to making such a significant and sudden change when, as I mentioned in Part 2, he said that **'All great truths begin as blasphemies'**—and, indeed, there's no greater and more sudden a change than when the redeeming and denial-destroying understanding of the human condition arrives!

The question this all raises is how are we to manage the arrival of the liberating but at the same time all-exposing **'day of judgment'**, **'the day of reckoning'** when **'[Our] nakedness will be exposed'** and **'not one secret of** [ours will be] **concealed'**? The simple answer is that we don't try to confront all the truth about our corrupted condition. Once we have investigated these understandings sufficiently to know that they have explained the human condition, we don't actually need to know any more than that to take up a life where we direct all our thoughts and energies into supporting these understandings, and to the repair of the world that they finally make possible. We don't *need* to know the full extent of the truth that this information reveals about the upset state of humanity; and we particularly don't need to know how it explains and reveals everything about our own human-condition-afflicted life. What did Bono say we would do when we have **'conquered...**[the] **past, [and] the future is here at last'**? He said **'The ruins to the right of me, will soon have lost sight of me'**. Yes, we don't dwell on the **'ruins'** of our corrupted condition. We get the truth up and we move on.

Furthermore, if you *do* become overly confronted by what is being presented, your resigned self's natural reaction could be to try to attack and deny it in order to protect yourself—to, in effect, reinstate all the denials, which means advocating humanity's retreat *back into* Plato's dark cave of denial that the human race has been desperately trying to find its way out of for 2 million years. The fact is, if you do become overly confronted, you could become defensive, angry and retaliatory toward the information, and the consequence of such a response could be to, in effect, sabotage the efforts of every human who has ever lived to bring the human race to this dreamed-of moment of its liberation. In short, <u>the effect of overly studying this information, studying it beyond your degree of security of self can cope with, can be both dangerous to you and dangerous to the human race, and no one should want, nor risk, either of those outcomes</u>.

WRONG RESPONSE   Attack the truth   Extinction of the human race
                                            from terminal alienation

RIGHT RESPONSE   Support the truth without   A transformed, human
                   overly confronting it   condition free world

Drawing by Jeremy Griffith © 1996-2014 Fedmex Pty Ltd

It makes sense that while the human race heals from its psychosis, a transition that will require a few generations, the more insecure should avoid overly confronting and studying the human condition and entrust that responsibility to those who are more sound. Of course, the effect of this sensible arrangement will be, as it now needs to be, that the more sound, least upset, more soul-connected, least alienated will be the ones who lead the human race back home to health and happiness. In the past our instinctive self or soul has been repressed because of its unjust condemnation of our intellect's search for knowledge, but at last that devastating state of soul separation/alienation/denial is now over. So instead of the emphasis being on the intelligent and the intellectual, it changes to being on the soulful and instinctual— from IQ to SQ. This healthy change of emphasis was perfectly anticipated by Christ when he spoke of a time when **'the meek... inherit the earth'**, and **'many who are first will be last, and many who are last will be first'**, and **'The stone the builders rejected has become the capstone'**—and by Bob Dylan in his 1964 song *The Times They Are A-Changin'*, **'The slow one now will later be fast / As the present now will later be past / The order is rapidly fadin' / And the first one now will later be last / For the times they are a-changin'.'** Remember, this emphasis on soul now is not a new form of elitism because the whole idea of good and bad, superior and inferior, has been eliminated with understanding that upset is a heroic, worthy and good state. (pars 1186-1188)

It is true that by not fully confronting the extent of the upset within yourself and avoiding looking too deeply into all the truth about the human condition, you are practising some denial/dishonesty, but by living in support of the compassionate full truth about the upset state of the human condition you are, in the greater scheme of things, ensuring the truth is being fully established and denial fully eliminated.

All that really matters now is that the truth is kept alive and that it is disseminated to the world's population, because it alone can heal the human race and save the world. All everyone should do now is support the truth about the human condition and it will achieve everything everyone has ever dreamt of. If we look after this information it, in turn, will look after each of us and the world. That is the mantra of the new world that understanding of the human condition brings about.

As I mentioned at the end of Part 2 when I was talking about the 'judgment day' shock that accompanies the arrival of the truth

<u>about the human condition, it is the enormous relief and joy of being able to free ourselves from the human condition that helps us most of all to deal with the sudden arrival of understanding of the human condition.</u> As I said, the procrastinating stage that <u>is characterised by the psychological wrestling match of 'I don't want to face the truth, I'm in deep shock, this is too big an upheaval and change to my life, it's all an unbearable blasphemy', won't last long once the wonderfully liberating potential of these understandings catches on</u>.

So while the psychological upset within humans has not been eliminated, because again that is a process that will take a few generations, in this new transformed way of living there is very little dishonesty, no irresponsibility, no weakness, no delusion, no deity, no worship, no focus on a personality, no faith, no dogma, and no guilt. What we have now is so relieving and *so exciting* that when this way of living catches on it *will* sweep the world!

───────────────

IN SUMMARY, now that we understand the very good reason why humans had to set out in search of knowledge and defy our original ideal-behaviour-insisting instinctive self or soul, *all* the upset anger, egocentricity and alienation that unavoidably resulted from being unjustly criticised by our instincts is now rendered obsolete, unnecessary and meaningless. <u>No longer do we have to retaliate against our instinct's unjust criticism of our search for knowledge. And no longer do we have to retaliate against criticism of our subsequent upset/corrupted state</u> because our upset/corrupted state has been defended with truthful, compassionate understanding at the most profound level. And <u>no longer do we have to try to prove our worth</u> because our worth has been established at the most fundamental level. And <u>no longer do we have to deny any confronting truths about our immensely upset/corrupted condition</u> because no longer are there any truths about our upset/corrupted state that condemn us. Our upset lives are now explained and defended, which means <u>we no longer have to be preoccupied compensating for that upset by finding forms of self-aggrandisement, by seeking self-distraction, or by chasing relief through materialistic forms of compensation for all the hurt we experienced growing up in an immensely human-condition-afflicted world</u>. And <u>nor do we have to delude ourselves that we are an upset-free, ideally behaved person by taking up support of a pseudo idealistic cause</u>. In other words, we no longer need to *artificially* make ourselves feel good about ourselves because our goodness has now been established at the deepest, most profound, *real* level.

And, most wonderfully, what happens when we humans give up our old ways of *coping* with the human condition and take up the new way of *living* that understanding of the human condition has made possible, is we naturally transition to a genuinely and authentically cooperative and selflessly behaved person, a truly integrative part of humanity. Even though we are not yet free of the psychologically upset state of our own personal human condition, we *can immediately* have a change of attitude and decide not to live out that upset state that remains within us. The overall effect in our lives is that, despite our retention of the upset state of the human condition, we are effectively free from its hold and its influence, which is an absolutely fabulous *transformation* to have made in an instant—in one simple decision!

You and every other human *can*, as it were, put the issue of all our upsets/corruptions in a 'suitcase', attach a label to it saying, 'Everything in here is now explained and defended', and simply leave that suitcase behind at the entrance to what we in the World Transformation Movement (WTM) call the Sunshine Highway and set out unencumbered by all those upset behaviours into a new world that is effectively free of the human condition. You can '<u>join the sunshine army on the sunshine highway to the world in sunshine</u>'! All the egocentric, embattled posturing to get a win out of life, all the strategising every minute of every day to try to find a way to compensate for feeling inadequate or imperfect or bad about ourselves, can suddenly end. The human race can leave Plato's dark cave where it has been hiding to escape the glare of the truth about its seemingly imperfect condition. And when we simply leave our whole 'must-prove-and-artificially-maintain-our-sense-of-worth, attack-and-deny-any-criticism' way of living behind as obsolete, all our thoughts and energies can finally be redirected into supporting and disseminating these human-race-saving understandings, and to repairing the world from all the damage our species' upset behaviour has caused— because with the human condition solved, all the upset that is causing the destruction of the planet can now end, which means it is at last possible to *properly* and permanently repair our environment. Excitement and meaning—based on liberating, truthful, honest understanding of ourselves and our world—is what we have to sustain ourselves now.

The immense excitement and relief of being effectively free of the human condition—the joy and happiness of being liberated from the burden of our insecurities, self-preoccupations and devious strategising; the awesome meaning and power of finally being genuinely aligned with the truth and actually participating in the magic true world; the wonderful empathy and equality of goodness and fellowship that understanding of the human condition now allows us to feel for our fellow humans; the freedom now to effectively focus on repairing the world; and, above all, the radiant aliveness from the optimism that comes with knowing our species' march through hell has finally ended and that a human-condition-free new world is coming—CAN NOW TRANSFORM EVERY HUMAN AND THUS THE WORLD.

From being a human-condition oppressed and depressed alienated person, you and all other humans can now be TRANSFORMED into redeemed, liberated-from-the-human-condition, exhilarated, ecstatic, enthralled-with-existence, empowered, world-transforming LIFEFORCES. This exhilarated, ecstatic, enthralled-with-existence aspect is the 'Life' in 'Lifeforce', while the empowered, world-transforming aspect is the 'force' in 'Lifeforce', so LIFEFORCE covers both the personal benefit and the benefit to the world in one word. From being human-condition-stricken, Plato's-dark-cave-dwelling, effectively dead humans, we become <u>Transformed Lifeforces</u> or simply <u>LIFEFORCES</u>! *That* is the difference the arrival of understanding of the human condition makes to the human race!

To confirm all that has been said, I recommend you watch some short videos in which WTM members describe their own transformation to a fabulous life in the new human-condition-understood world at www.humancondition.com/video-transformation.

Again, chapter 9 of *FREEDOM* contains the full description of how understanding of the human condition is able to immediately transform your life, and the lives of all humans.

Finally, as mentioned in the Notes to the Reader, while copies of *Transform Your Life And Save The World* and *FREEDOM* can be purchased from bookshops (including Amazon), what they are presenting is so important they are, and will always remain, *freely* available to read, print or share at our website below.

So please introduce others to this understanding of the human condition by telling them they can immediately, easily, and for no charge, read this little book, because the explanation it contains is the *only* thing that can save the human race from unthinkably horrific terminal levels of writhing and deadening alienation/soul-separation/psychosis.

www.HumanCondition.com

Appendix

Summary of our species' heroic journey
from ignorance to enlightenment

Now that we can understand the good reason we conscious humans corrupted our all-loving original instinctive self or soul, and can therefore truthfully acknowledge all the stages our ancestors went through, from our innocent, early bonobo-like *Australopithecus* childhood ancestors through to the upset adolescent *Homo* varieties of our ancestors, the true story of our species' journey from ignorance to enlightenment can finally be told. This journey is described in chapter 8 of *FREEDOM* where the following sequence of fossil skulls of our ancestors appears, as well as the series of drawings I created (and which appear throughout chapter 8) to illustrate the various stages I describe the human race as progressing through (see overleaf).

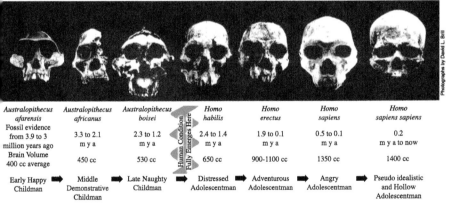

Photographs by David L. Brill

Australopithecus afarensis Fossil evidence from 3.9 to 3 million years ago Brain Volume 400 cc average	*Australopithecus africanus* 3.3 to 2.1 m y a 450 cc	*Australopithecus boisei* 2.3 to 1.2 m y a 530 cc	Human Condition Fully Emerges Here	*Homo habilis* 2.4 to 1.4 m y a 650 cc	*Homo erectus* 1.9 to 0.1 m y a 900-1100 cc	*Homo sapiens* 0.5 to 0.1 m y a 1350 cc	*Homo sapiens sapiens* 0.2 m y a to now 1400 cc
Early Happy Childman ➡	Middle Demonstrative Childman ➡	Late Naughty Childman		Distressed Adolescentman ➡	Adventurous Adolescentman ➡	Angry Adolescentman ➡	Pseudo idealistic and Hollow Adolescentman

The Stages in Humanity's Journey
From chapter 8 of *FREEDOM: The End Of The Human Condition*

INFANCY
When nurturing developed our moral soul

Bonobos Matata and her adopted son, Kanzi

CHILDHOOD
When consciousness developed

EARLY HAPPY INNOCENT CHILDMAN

You're bad!

(You should share them)

I feel hungry: why not take the apples? An innocent mistake made by a mind trying to reason how to behave.

Our moral conscience

Our mistake-prone conscious mind

MIDDLE DEMONSTRATIVE CHILDMAN

You're bad!

Our moral conscience

'Why shouldn't I push Johnny over in the playground?' –the frustrated, naughty, bullying final stage of childhood

LATE NAUGHTY CHILDMAN

ADOLESCENCE

When we searched for our identity, particularly
for why we weren't ideally behaved

Detail of Francis Bacon's honest painting
of the alienated human condition

Deluded, alienated 'Adolescentman'

You're bad!

Our moral conscience

EARLY SOBERED ADOLESCENTMAN

You're bad!

Our moral
conscience

Approaching Resignation

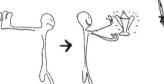

The moment of Resignation

The selfish, power-fame-fortune-
and-glory seeking resigned adult

DISTRESSED ADOLESCENTMAN

Now you're
really bad!

Our moral conscience

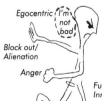

Egocentric I'm not bad!

Block out/
Alienation

Anger

Escape

Fuck/Destroy
Innocence

Materialism like
fast cars and
swimming pools
serviced our
escape and
egotism

ADVENTUROUS ADOLESCENTMAN

ADOLESCENCE continued

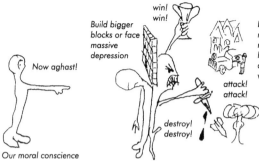

ANGRY ADOLESCENTMAN

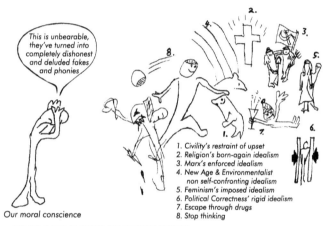

1. Civility's restraint of upset
2. Religion's born-again idealism
3. Marx's enforced idealism
4. New Age & Environmentalist non self-confronting idealism
5. Feminism's imposed idealism
6. Political Correctness' rigid idealism
7. Escape through drugs
8. Stop thinking

PSEUDO IDEALISTIC ADOLESCENTMAN – the massively deluded, 'I'm good and this is the way to solve the world's problems', weak abandoner of the battle

HOLLOW ADOLESCENTMAN

ADULTHOOD
When we found understanding of ourselves

Old World   New World

RECONCILING UNDERSTANDING OF THE HUMAN CONDITION FOUND

*Our now baseless fear of exposure
of our corrupted human condition*

*Resistance to letting go our
now-obsoleted selfish power,
fame, fortune and glory-seeking
way of justifying our worth*

THE PROCRASTINATION STAGE

THE TRANSFORMED LIFEFORCE STATE

Lightning Source UK Ltd.
Milton Keynes UK
UKOW07f1845230916

283669UK00013B/102/P